Civil 3D and AutoCAD
Professional Tips and Techniques

Topic-based learning for intermediate and advanced users
Recommended for Civil Engineers

- Learning through Q&A
- Useful tips and tricks
- Focus on real-world projects
- Stripped of generalities and theory
- Highlight frequent questions, problems, and errors
- Include practical examples for every topic
- Provide instructive illustrations and diagrams for every topic
- Can serve as a Civil 3D encyclopedia for learners of all stages

Road construction
Cartography
Transmission lines
Land leveling
Land subdivision
Special points and issues

Author: Javad Noormohammadi

Table of contents

Introduction

This handbook enhances the ability of cartographers and civil engineers to work with the powerful Softwares Civil 3D and AutoCAD by providing practical illustrative examples and problem instances. Here, we try to avoid the theoretical and general description of software features and capabilities and instead focus on the practical use of features on a topical basis. The educational materials are organized for step by step learning, but the handbook also serves as a compact encyclopedia allowing the reader to access the content of interest directly. The handbook is designed to meet the shared needs of users, answer the frequently asked questions, and resolve the problems commonly encountered when using Civil 3D.

We hope that the reader will find this document helpful and informative.

Tip:

To receive the attached files, email us at javad_nm_aus@yahoo.com

1- How to apply elevation adjustments to imported points without changing the elevations in the main file.

Suppose you want to change the elevation of all surveyed features (e.g. the corner of the buildings) by a certain amount without changing the elevation of points in the main file, that is, just by applying adjustments as you import them into the software.

For this exercise, open the folder points files and then the file named elevation adjustmentB.txt. In this file, the first four points are marked with the code *building*. Now, suppose you want to import these points into the software with their elevation reduced by for example 40cm. To do so, just type a comma after the code (here, *building)* followed by the amount of change you want to apply. Finally, save the file as elevation adjustment-2.txt.

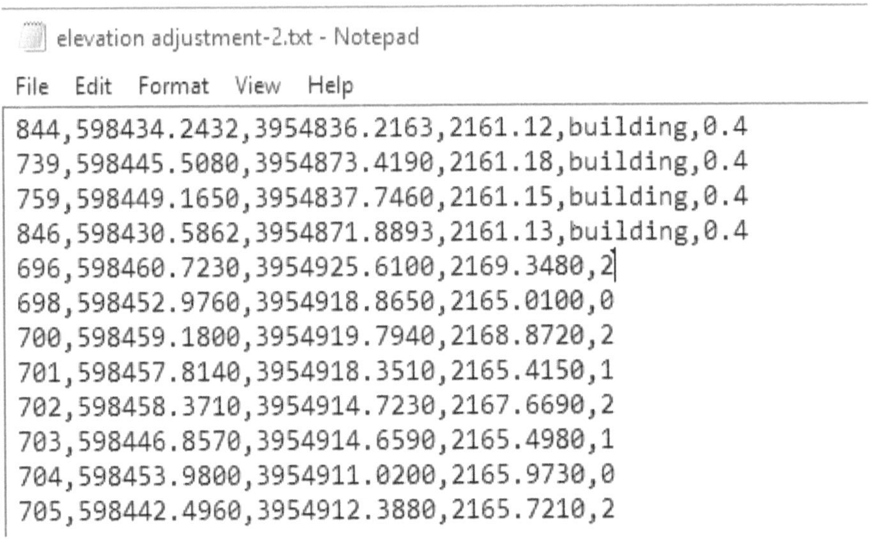

---

 elevation adjustment-2.txt - Notepad

File   Edit   Format   View   Help

```
844,598434.2432,3954836.2163,2161.12,building,0.4
739,598445.5080,3954873.4190,2161.18,building,0.4
759,598449.1650,3954837.7460,2161.15,building,0.4
846,598430.5862,3954871.8893,2161.13,building,0.4
696,598460.7230,3954925.6100,2169.3480,2
698,598452.9760,3954918.8650,2165.0100,0
700,598459.1800,3954919.7940,2168.8720,2
701,598457.8140,3954918.3510,2165.4150,1
702,598458.3710,3954914.7230,2167.6690,2
703,598446.8570,3954914.6590,2165.4980,1
704,598453.9800,3954911.0200,2165.9730,0
705,598442.4960,3954912.3880,2165.7210,2
```

Figure B2

To apply these adjustments during the importing process, you need to create a point file with proper elevation adjustment format. To do this, when importing the points, click on the *Manage format* button to the right of the box named *Specify point file format* (Figure B3). In the opened window, click on the *New* button, select *User Point,* and press *OK*.

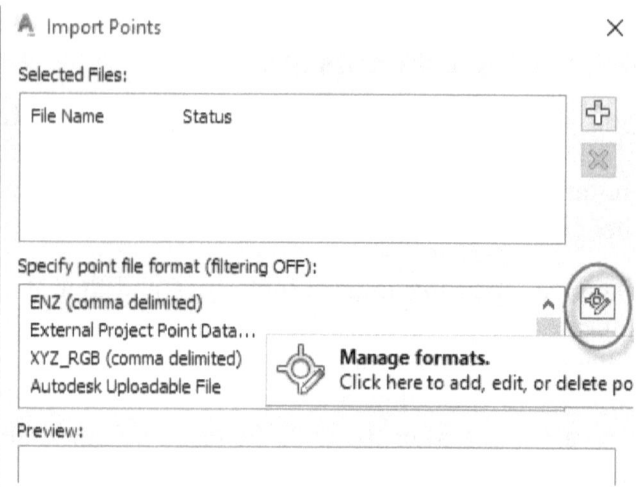

Figure B3

This will open a window named *Point File Format* (Figure B4), where you must give a name to your format and then set the *Format Options* to *Delimited* and type a comma in the box to its right.

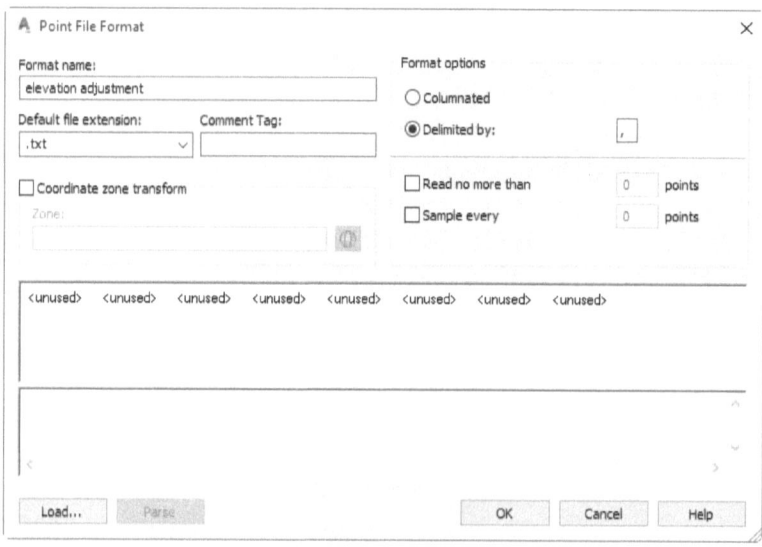

Figure B4

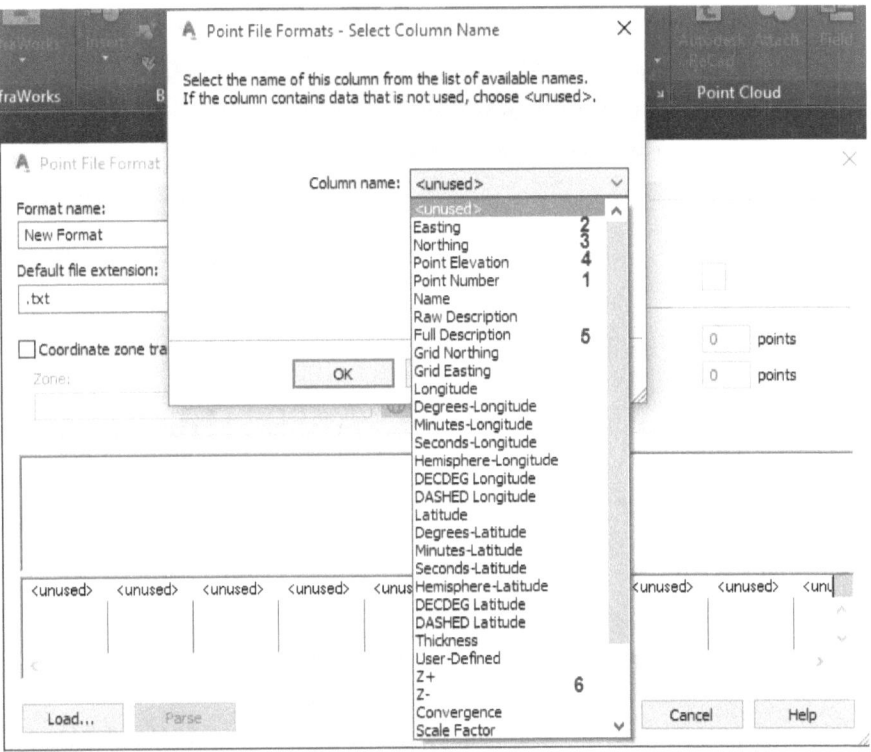

Figure B5

Displayed in the bottom of this window (Figure B5) are the column headers, which are marked as *unused* and can be clicked on to specify the desired data type. To import the points with desired adjustments, you must create *a PENZDZ*-file format. To do this, click on the first *unused* column and select *Point Number* from the list. Similarly, select *Easting* option for the second column, *Northing* for the third column, *Point Elevation* for the fourth column, *Full Description* for the fifth column, and *Z* for the sixth column (see steps 1 to 6 in Figure B5). Alternatively, you can increase the elevation of points by selecting *Z+* for the last column. Press *OK* to create the custom format.

After creating the format, select it from the list of formats and then select the file <u>adjustment-2.txt</u> for import. Finally, check the box *Do elevation adjustment if possible* so that the software applies the specified adjustment as it imports the points.

2- How to make X-Y coordinates appear beside points?

This section explains how you can make the software display X-Y coordinates near points (as shown in Figure B6). For this purpose, you need to create a *Point Label Style* with the right settings to display X-Y coordinates.

X=598414. 77
Y=3954821. 49

X=598416. 70
Y=3954821. 14

X=598415. 62
Y=3954820. 42

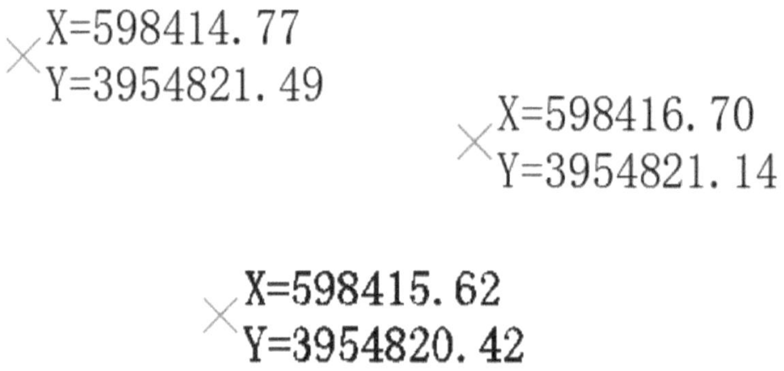

Figure B6

To do so, in the *TOOLSPACE* tab, right-click on the point group of your choice and select *Properties* to open the *Point Group Properties* window (Note: If you do not see the *TOOLSPACE* tab, activate it in the *View* tab). In this window, open the *Point Label Style* menu (Figure B7) and select *Create New* to make a new label style (alternatively, you can choose one of the existing *Label styles* and press *Edit Drawing Selection*). This opens the *Label Style Composer* window shown in Figure B8.

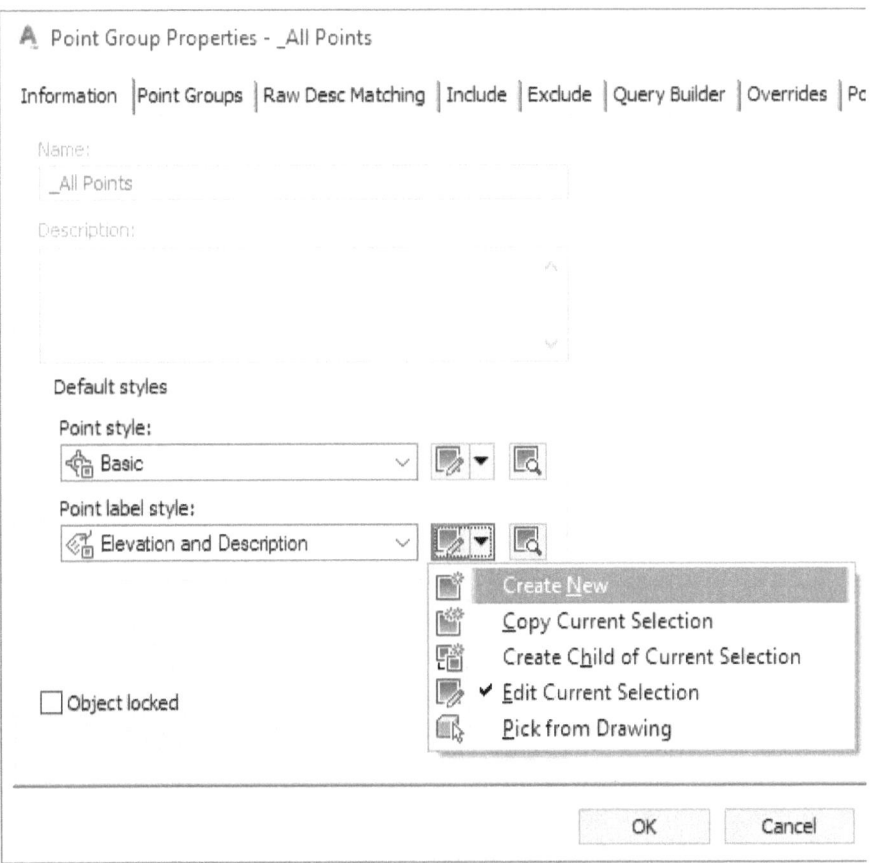

Figure B7

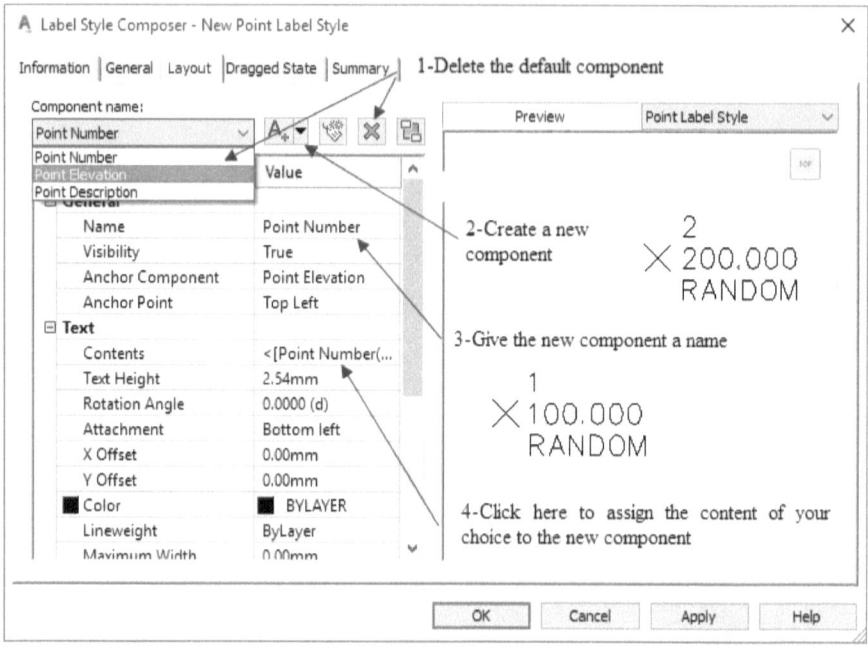

Figure B8

In the Information tab of this window, you must name your label. Then open the *Layout* tab and delete the default components (Point Number, Point Description, Point Elevation) by selecting them and clicking on the red X button.

After deleting the components, click on the button shown in Figure B8 and select *Text* (this is the format of information in the new component). After typing a name in the *Name* field, click on *Contents* to open a window where you can define the content of your choice. In this window (Figure B9), first, remove the *Label Text* (default option) from the pane on the right.

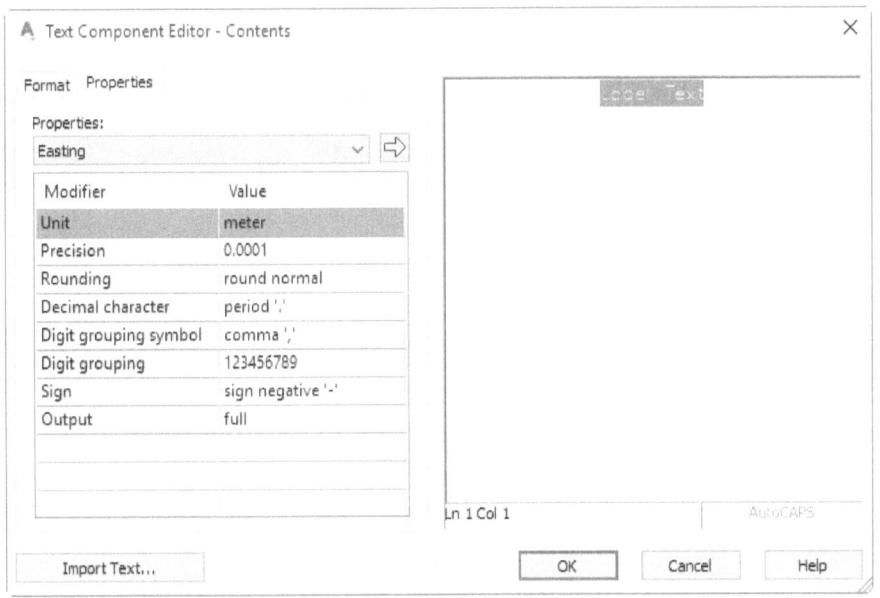

Figure B9

Now, if you want the display X-Y coordinates with a prefix and suffix, like

X= 125.36 m

Y= 185.63 m

type $X =$ in the left pane (this will be the prefix) and then select *Easting* option in *Properties* menu. After setting the related parameters (like the number of decimal places displayed), press the arrow key next to the *Properties* menu to move it to the front of the $X =$ in the right pane. Then simply type *m* in front of it (this makes *m* appear as a suffix) and finally press enter. Repeat the same process for Y value but this time use *Northing* option instead of *Easting* and type $Y =$ as the prefix. In the end, the result should be similar to the following:

X= <[Easting(Um|P2|RN|AP|GC|UN|Sn|OF)]> m

Y= <[Northing(Um|P2|RN|AP|GC|UN|Sn|OF)]> m

You can adjust the font settings, add special symbols, etc. on the *Format* tab. To change the settings, you just have to select the text displayed on the right and make the desired adjustments.

After completing the above steps, press *OK* to see the X-Y coordinates appear right next to the points.

3- How to extract the *Scale Factor* of points

Some projects may require you to apply a *Scale Factor* while surveying or setting out. In such projects, you can use Civil 3D to obtain the scale factor of survey stations without static GPS surveying. For this purpose, you must apply the coordinate system and the zone number of the area to the point file being imported. Also note that since the scale factor is a function of elevation, elevation should be from the mean sea level (using local elevations will cause a significant error).

After the above preparations, open the *Point* menu and click on *Utilities* and then *Geodetic Calculator* (Figure B12) to open the window shown in Figure B13.

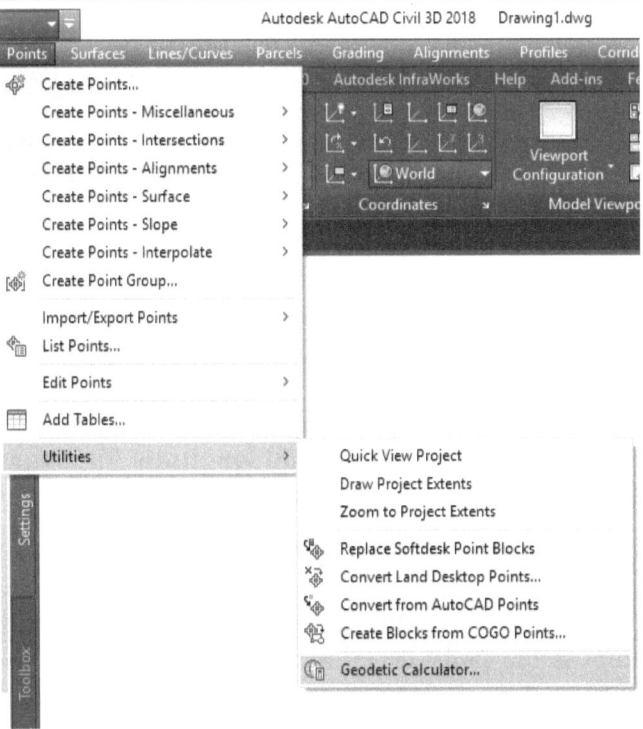

Figure B12

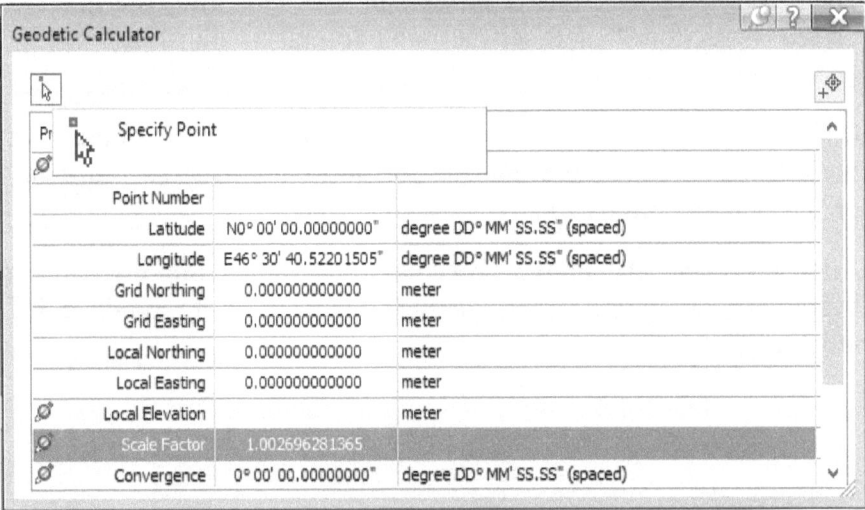

Figure B13

In this window, press the *Specify Point* button and then click on the point from which you want to extract the scale factor (the point coordinates will appear in the *Scale Factor* box).

Note: When defining the coordinate system, make sure that the length unit displayed in the *Drawing Unit* box is *Meter*, not *Feet*, otherwise there will be an error in the calculated scale factor.

To calculate the scale factor of a large number of survey stations and print their scale factors alongside their outputs, you have to create a new format (as explained in section 3) and add *Scale Factor* in the sixth column.

4. How to import points in latitude and longitude format and convert them into UTM coordinates

For this exercise, we use the file named lat-long.txt in the folder points file. In this file, the first column is the point number, the second column is latitude, the third column is longitude, the fourth column is elevation, and the fifth column is the point description or code. Latitude and longitude values are displayed to 8 decimal places. As previously explained, to import this point file, you need to first define the global coordinate system and zone number of your drawing file.

Then, you have to create a format for the point file (the method is described in section 2) and set the point format as illustrated in Figure B14.

If the latitudes and longitudes given in the point file have minutes and seconds listed in separate columns, you must use *Degrees-Latitude* or other appropriate options when defining the format.

After following the above procedure, the points imported into the software will be in UTM format and can be used to produce UTM coordinates and outputs.

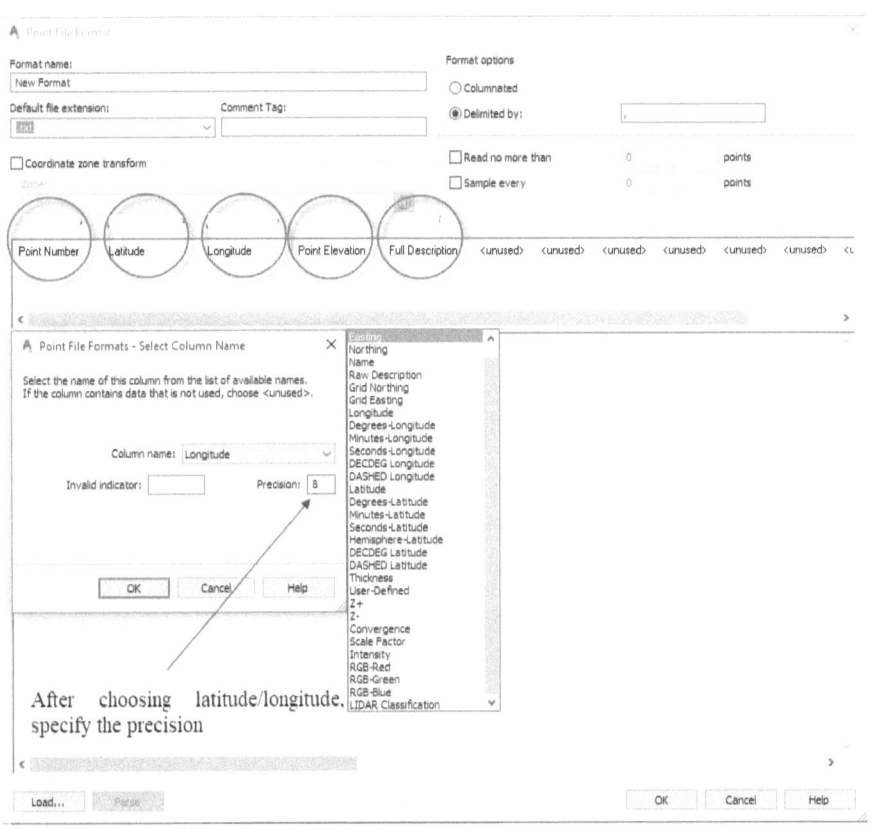

Figure B14

5- How to adjust the point label so that elevation's decimal point appears on the marker.

One of the standard formats for marking points in civil engineering plots is to place the elevation's decimal point on the point marker. But this format is not included in Civil 3D and must be manually created as a new point label.

To do this, after importing the points, go to the *TOOLSPACE* window and the *Prospector* tab, right-click on the point group of interest and select *Properties* (Figure B15). In the opened window, go to the *Information* tab and click on the arrow to the right of *the Point Style* box to open a cascading menu, and click on *Copy Current Selection*.

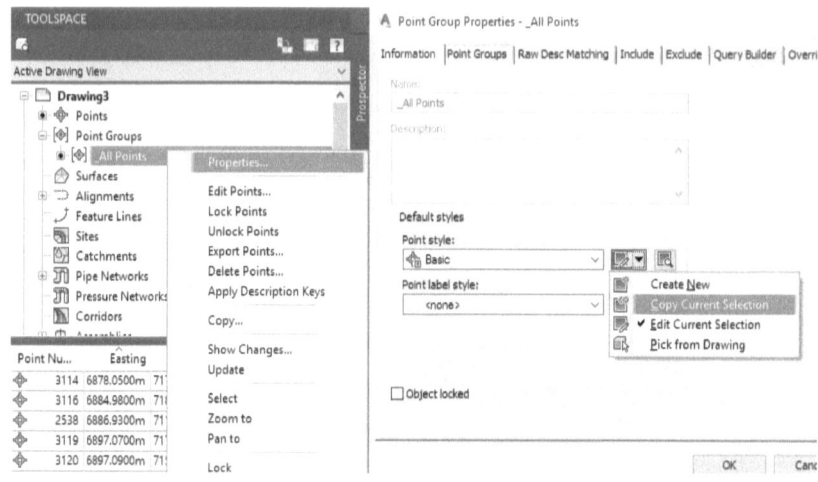

Figure B15

In the *Information* tab of the opened window (Figure B16), give your marker a name. Then go to the *Marker* tab, select the *Use custom marker* option, and then select the two markers shown in the figure. Type 0.1 in the *Size* box and press *OK*.

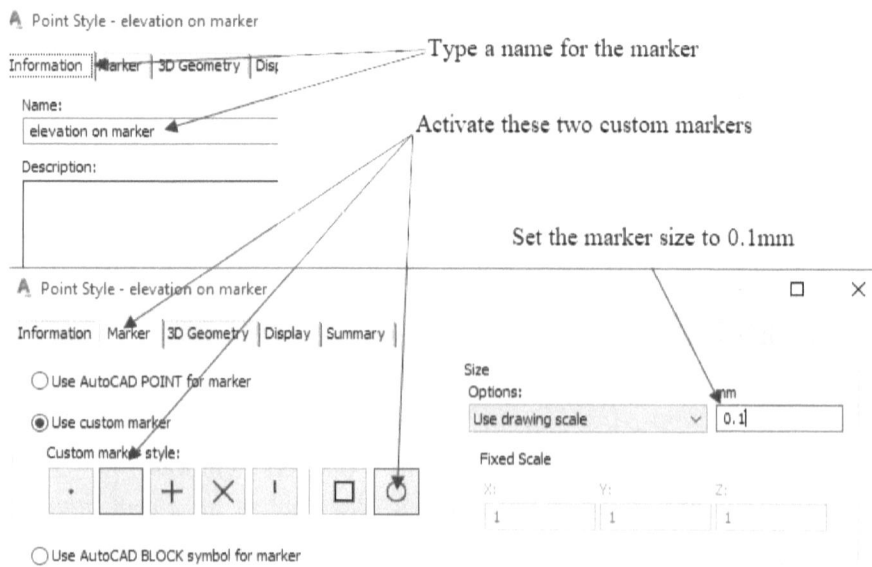

Figure B16

The next step is to adjust the point label. As shown in Figure B17, set the *Point Label Style* to *Elevation Only*, and then click on the arrow to its right and select *Copy Current Selection*.

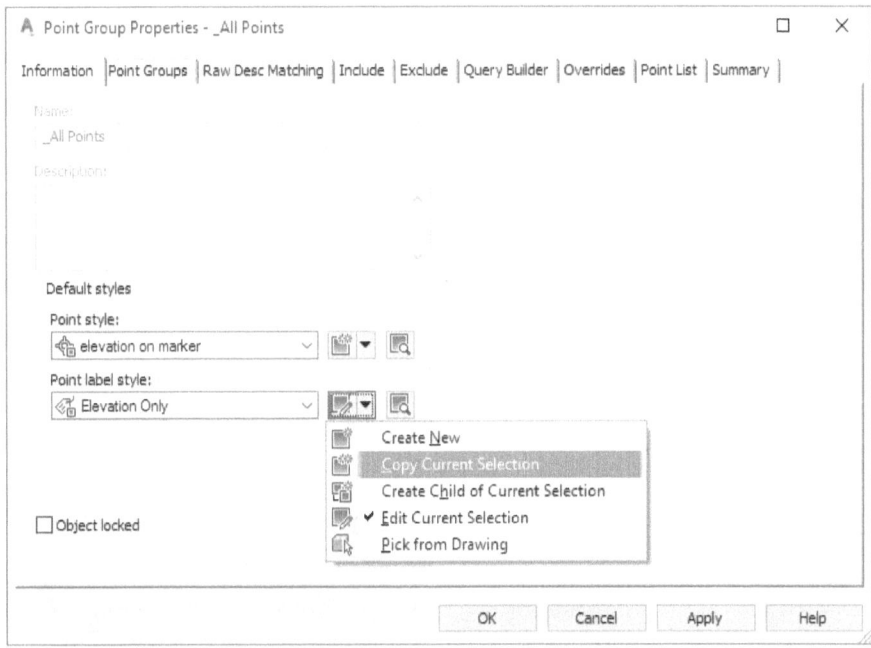

Figure B17

In the opened window, go to the *Information* tab and type a name for your label. Then, go to the *Layout* tab and delete the *Point Elevation* component by clicking on the red X button (Figure B18).

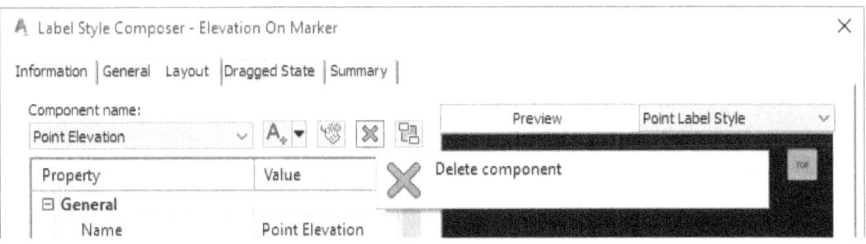

Figure B18

Now, as shown in Figure B19, you must use the *Create Text Component* button to create two components: one for the integers to be placed before the decimal point and another for the decimals to be placed after it. Enter the name of the new component in the *Name* box. Here, we named the first component <u>integer</u> and the second one <u>decimal</u>.

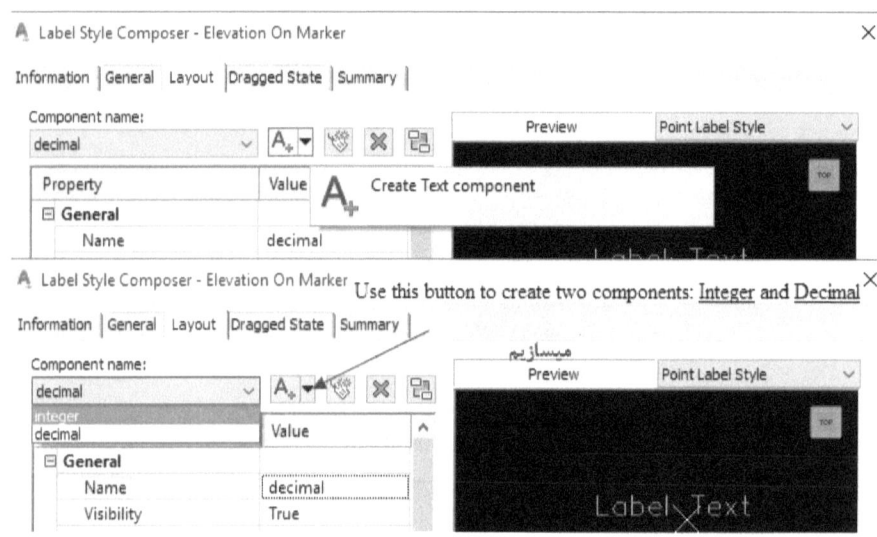

Figure B19

Once components are created, their contents must be specified. For this purpose, select the component named <u>Integer</u> and click on the box to the right of *Contents* (Figure B20).

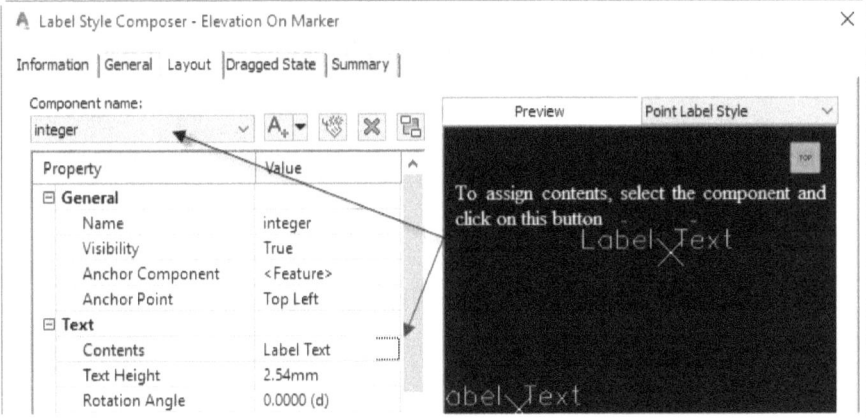

Figure B20

This opens the content specification window. In this window, first, select and delete the text in the right pane. Then, as shown in Figure B21, change the *Properties* to *Point Elevation*, set the *Precision* to 1, and set the *Output* to the *Left Of decimal*. Click on the arrow button to apply the changes and then press *OK* to close the window.

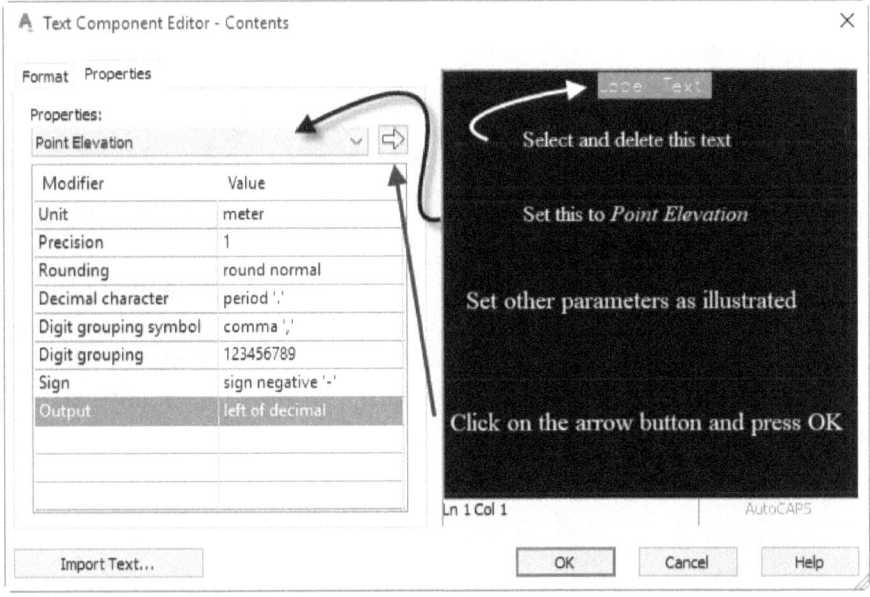

Figure B21

You need to repeat this procedure for the component named Decimal, with the difference that the *Precision* variable should be set to 2, and the *Output* should be set to the *Right Of Decimal*.

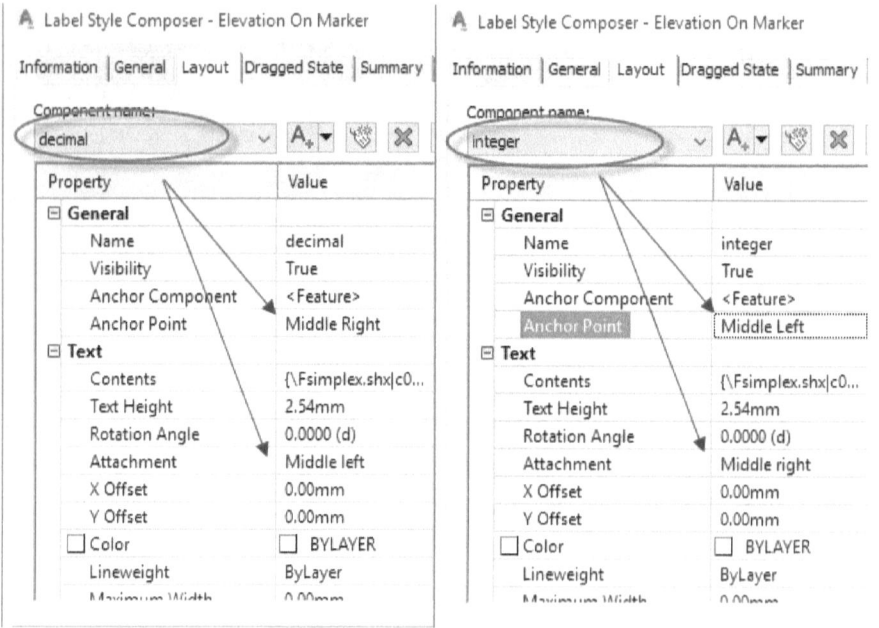

Figure B22

The next step is to adjust the location of these components. As shown in Figure B22, for the component Integer, set the *Anchor point* to *Middle Left* and set the *Attachment* to *Middle Right*. For the component Decimal, you have to set the *Anchor point* to *Middle Right* and set the *Attachment* to *Middle Left*. After pressing *OK*, your points will be displayed as shown in Figure B23, which is the standard format for many national cartography and survey organizations around the world.

2303·49

2299·76

Figure B23

6- How to save the customized labels and markers for use in other drawings?

As expected, many Civil 3D users prefer to save their templates, customizations, and settings for use in other projects. For example, when you create a style to display the elevation's decimal point on the marker and want to use it in multiple drawings, it is more sensible to export the style rather than rebuilding it over and over again. To do this, you must save the drawing in the *Template* folder in the *AutoCAD Drawing Template* format (Figure B24). Give the file a name that fits its purpose, such as <u>Elevation On Marker.dwt.</u> In your other project, run the *Layout* command on the command line to receive the response shown in Figure B25.

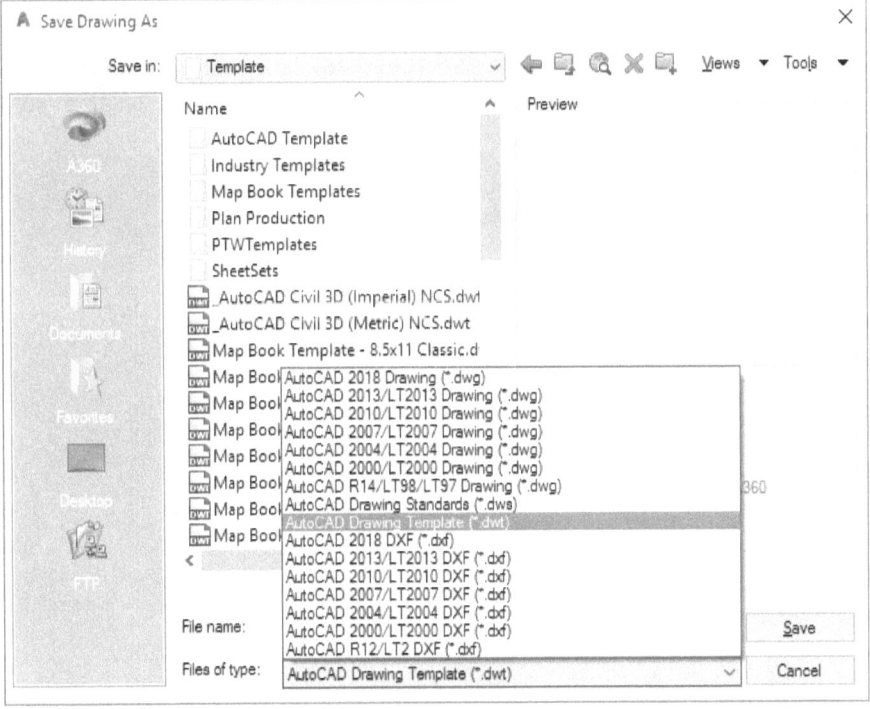

Figure B24

In response, type T and press enter to select the *Template* option. In the opened window, click on the saved template and select *Layout1* and press *OK* to apply the saved settings. Now, you can use the label created in your previous project in your current one.

Figure B25

7- How to add prefix/suffix to a point label?

Sometimes, you need to adjust the point label to include a prefix (such as $X =$ or $Y =$) or suffix (such as m) in order to clarify the nature or unit of the displayed information. For example, suppose you want the point elevation to be displayed as shown in Figure B26. To do this, you need to go to the *TOOLSPACE* window and then the *Prospector* tab, right-click on the point group whose label you want to customize and select *Properties*.

Figure B26

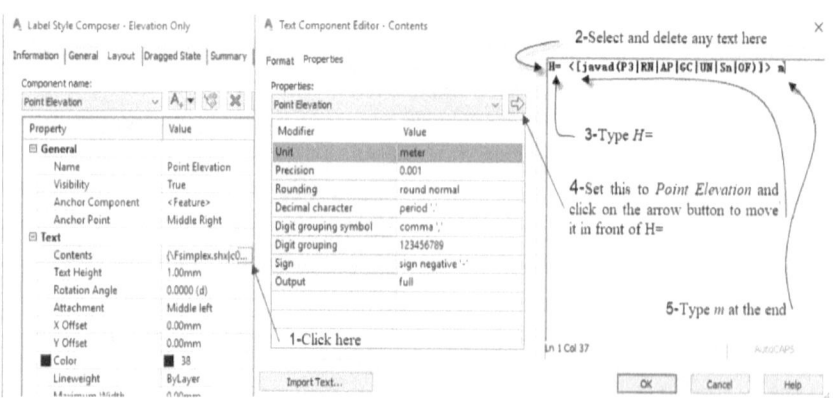

Figure B27

In the opened window, set the *Point Label Style* to *Elevation Only* and then open the cascading menu to its right and select *Edit Current Selection*.

In the opened window (Figure B27), go to the *Layout* tab and click on *Contents* to open the window shown on the right side of Figure B27. In this window, delete the contents of the right pane and type *H* =, open the drop-down menu below *Properties* and select *Point Elevation*, click on the arrow to insert the parameters in front of *H* =, and finally type *m* in front of inserted parameters.

8- How to color the points based on their elevation?

In many projects, it is technically or aesthetically recommended for the points in a specific elevation range to have the same color or the points in different elevation ranges to have distinguishable colors.

For this exercise, import a topographic map named 2 Elevation from the Points File folder. The supposed goal here is to implement an irrigation scheme consisting of building water tanks in areas with an elevation of more than 1420 meters. Thus, the first step is to identify the points with an elevation of 1420 meters on the topographic map. It will also be more convenient if the points of interest (the ones with an elevation of more than 1420 meters) are distinguishable by color. To make this happen, go to the *Point* menu and select *Create Point Group* to open the *Point Group Properties* window (Figure B35).

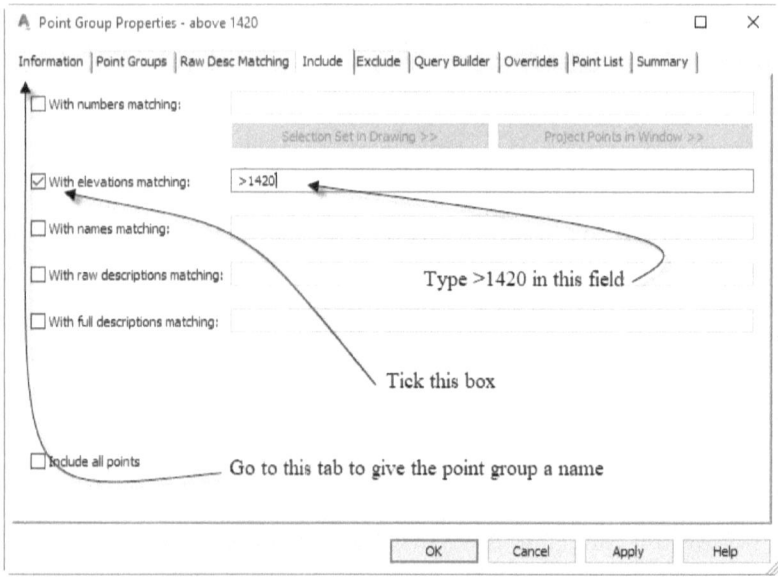

Figure B35

In the *information* tab of this window, type a name for your point group. Then go to the *Include* tab, tick the *With elevation matching* option and type >1420 in its field, and finally press *OK*. You just created a point group consisting of all points with an elevation of above 1420 meters. Now, to make these points easily distinguishable from others, go to the *TOOLSPACE* window and then the *Prospector* tab, right-click on the point group you just created, click on *Properties*, and change the color of the marker or text as needed (Figure B36).

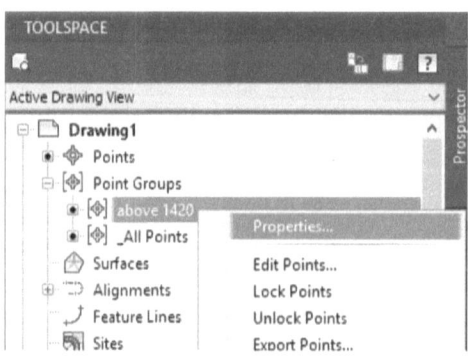

Figure B36

9- How to remove the points of a specific elevation range?

Suppose that in the previous exercise, you want to remove the points with elevations of between 1410 and 1420. To do this, create a point group as explained in the previous exercise, but after checking *With elevation matching* option, type 1410B420 in its field. Then, as shown in Figure B37, click on the created point group and select *Delete Point*.

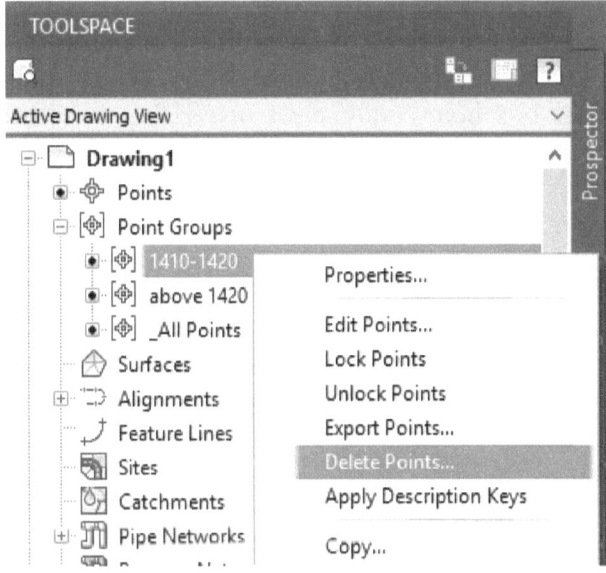

Figure B37

10- How to remove the points with a specific elevation range and a specific point code?

Suppose that in a topographic survey of an urban aura, you have measured the elevation of the bottom corners of multiple buildings, but because of field obstacles, you have had to make a few measurements of top corners as well. Now, you must remove these points to prevent error in the subsequent operations on the ground points. Thus, the goal here is to remove the points with the *BLD* code whose elevation is higher than the average of the region.

To do this, a conditional command must be issued to identify the points that meet both conditions of (i) having a *BLD* code, and (ii) having an elevation of more than the average of the area.

For this exercise, open the file bld.txt from the Points File folder. As you can see, this file contains several points with the codes *bld* and *topo*. You can also see that most of the *bld* points have an elevation of less than 1821.00, but there are several points of both *bld* and *topo* type that are located above the 1825.00 elevation level. To remove the points that have a *bld* code AND an elevation of more than 1820.00, go to the *Points* menu and select *Create Points Group*. After typing a name in the *Information* tab, go to the *Query Builder* tab and check the *Modify query* box. Then Right-click in the box below and select *Insert Row* to create a new row and write a conditional command as shown in Figure B38. Once the group is created, you can delete all the point in this group as explained in the previous exercise.

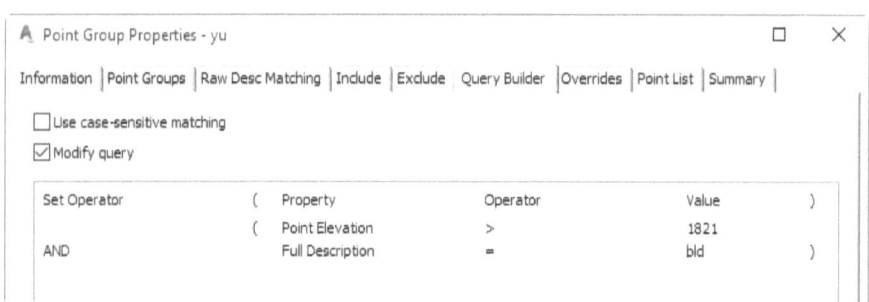

Figure B38

11- How to create a point group by mouse clicking?

To do this, go to the *Points* menu, click on *Create Points Group* option, type a name in the *Information* tab, and then go to *Include* tab. In this tab, tick the box named *With numbers matching* and then click on the *Selection Set in Drawing* button to the right (Figure B39). Select the points you want to include in the group and press *OK*.

Figure B39

12- How to group the points based on their code?

In highly detailed and busy maps, it is extremely convenient and rewarding to group the points based on their code so that they can be turned on and off as needed or given specific colors and labels according to their nature. For this exercise, import the file named Code-Point.txt from the Points File folder.

As you can see, this file contains 3508 points with multiple point codes, which if not grouped based on the code, will be very difficult to comprehend and work on. The codes present in this file include *MARZ*, *CHAH*, *BLD*, *FENCE*, *JADEH*, *JOOY*, *ROOD*, *TABLO*, *TIR*, and *TOPO*. The goal is to create, for each code, a point group of the same name and then place every point of that code in that group. To do so, select *Create Point Group* from the *Points* menu. In the *information* tab, enter the name of the code (e.g. *TIR*) in the *Name* field. Then go to the *Include* tab, tick the box *With full description matching*, type the name of the code (here *TIR*) in its field (Figure B40), and press *OK*. Repeat this process for all other codes.

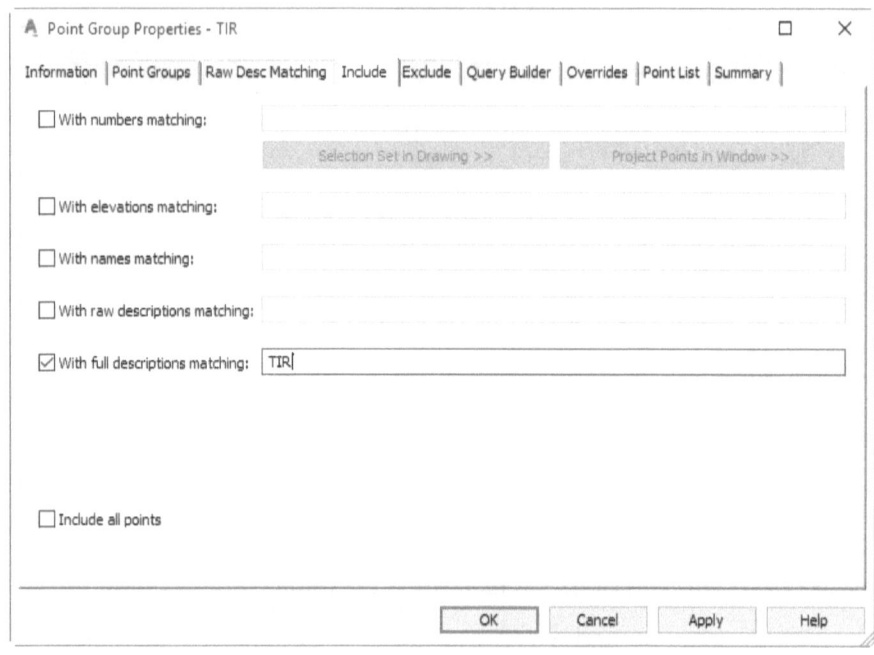

Figure B40

13- How to create a group of all the points whose code include a specific letter or word.

For this exercise, import the file named <u>Point bld.txt</u>, which comprises the points surveyed from an area with 13 buildings. In this file, buildings are given the code *BLD* followed by a number that specifies the exact building to which the point belongs, so the codes range from *BLD1* to *BLD13*. Using the method of the previous exercise, you can create 13 groups each dedicated to one code. But if you want to place all of the building points from *BLD1* to *BLD13* in a single group, you just have to tick the *With full description matching* box and type *BLD** in its field (Figure B41). After pressing *OK*, all the points whose code starts with the word *BLD* will be placed in the created group.

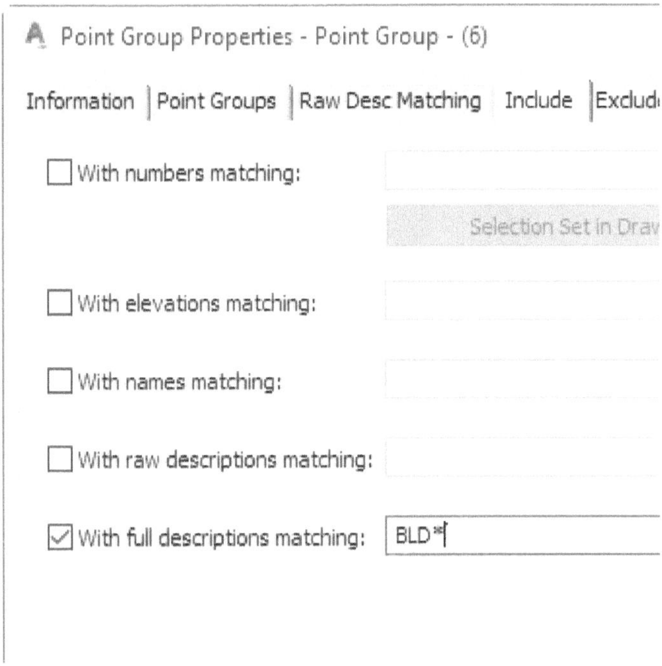

Figure B41

14- How to select all the points whose code include a specific letter or word without creating a point group.

Suppose that in the previous exercise, we want to select the points whose code begins with *BLD* not to create a group but to perform another operation such as deleting, moving, code editing, elevation editing, etc. For this purpose, type QSELECT in the command line and press enter. This command opens the *Quick Select* window shown in Figure B42. In this window, set the *Object Type* to *Cogo Point*, set the *Properties* to *Full Description*, set the *Operator* to *Wildcard Match*, and type *BLD** in the *Value* field. After pressing *OK*, this operator selects all the points whose code starts with *BLD*.

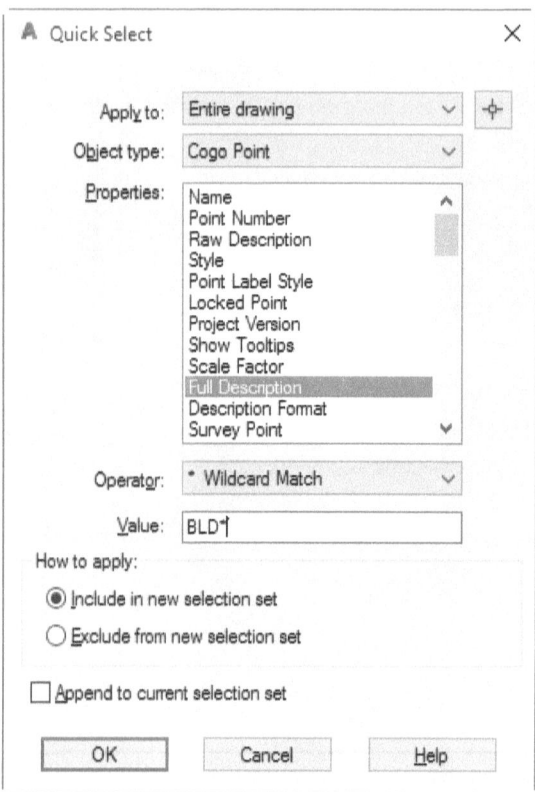

Figure B42

15- How to create a group of all the points whose code include a number.

For this exercise, import the file named <u>Number-in-code.txt.</u> In this file, some buildings are given codes such as 1BLD, BLD3, 2BLD6, BL5D, BLD12, BL16D, etc., but the codes of other features do not include any number. Thus, to create a group specifically for building points, you just have to select all the points whose code include a number.

To do this, start creating a point group as explained in the previous exercises. But this time in the *Include* tab, check the box next to *With full description matching* and type *1*, *2*, *3*, *4*, *5*, *6*, *7*, *8*, *9*, *0* (Figure B43).

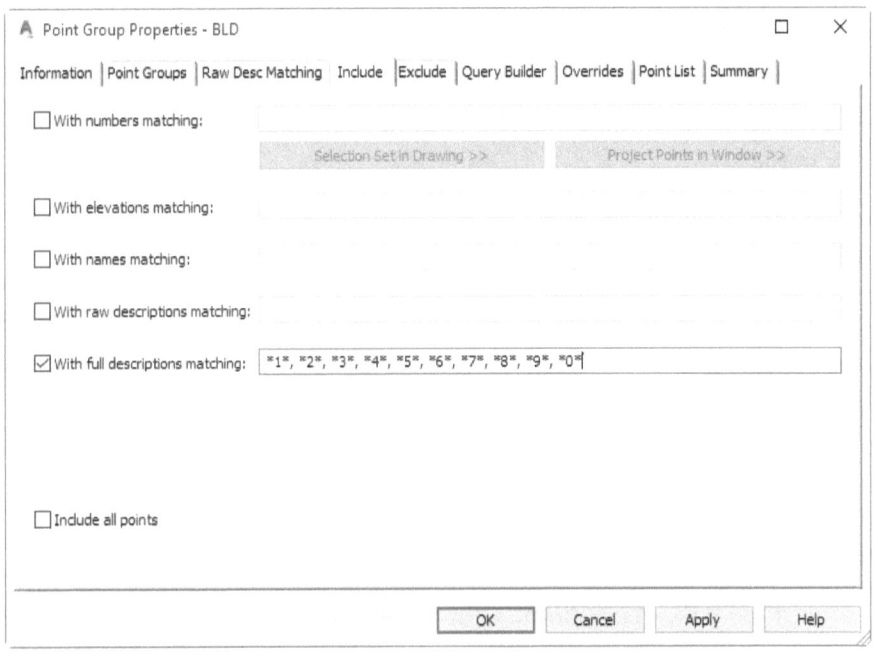

Figure B43

16- Description of characters for complex point selection

In exercises 16, 17 and 18, you learned how to create a point group or select a group of points based on certain combinations of letters, words, and numbers in their codes. Here, we are going to introduce the characters such as * and # that help you master this ability.

*

This character represents all and every letter, number, and symbol. For example:

If you type T* in the point selection field, every point whose code starts with the letter T (e.g. T1, TOPO, T-2) will be selected.

If you type *T, every point whose code ends with the letter T (e.g. 1T, ROOT, 2-T) will be selected.

And if you type *T*, every point whose code include the letter T will be selected.

# #

This character represents the numbers, and the number of times it is repeated represents the number of number characters that we search for. For example:

If you type T#, every point with the code T0, T1, T2, ... , T9 will be selected.

If you type T##, every point with the code T00, T01, ... , T99 will be selected.

If you type #T, every point with the code 0T, 1T, 2T, ... , 9T will be selected.

If you type JOOY#, every point with a code like JOOY1, JOOY2, ... will be selected.

If you type #J#, every point with a code like 1J9, 5J7, 9J0, ... will be selected

And if you type ##U#, every point with a code like 21U4, 10U5, ... will be selected.

@

This character represents letters, and the number of times it is repeated represents the number of letter characters. For example:

If you type @1, every point with the code A1, B1, ..., Z1 will be selected.

If you type @@1, every point with a code like AA1, AB1, BS1, ... will be selected.

If you type 1@, every point with the code 1A, 1B, ... 1Z will be selected.

And if you type @@9@@, every point with a code like AS9BD will be selected.

•

This character represents any character other than letters and numbers, and the number of times it is repeated represents the number of characters. For example:

If you type T., every point with a code like T=, T+, T- will be selected

If you type .T., every point with a code like +T=, =T=, -T+ will be selected

And if you type ..T., every point with a code like =+T- will be selected.

?

This character represents both letter and number. Unlike the character (*), this character allows you to specify the number of characters that you search for. For example.

If you type ?AB, every point with a code like 3AB, GAB will be selected.

But if you type ??AB, every point with a code like ASAB, 31AB, D3AB will be selected.

~

This character can be used for invert selection. In other words, the use of this character leads to the selection of any point other than what is specified by the succeeding argument. For example:

If you type ~*AB*, the points whose code include anything other than AB will be selected.

And if you type ~@1, the points with any code other than B1, C1, Z1, ... will be selected.

[...]

This can be used to define a specific set of characters for selection. For example:

If you type [AB]C, only the points with the code AC and BC will be selected.

And if you type [~AB]C, the software will select every point with a code like VC and DC, excluding AC and BC.

- in [...]

This can be used to define an interval of characters rather than specific characters. For example:

If you type [1-5]A, the software will select the points with the codes 1A, 2A, 3A, 4A, 5A, but not 6A or 0A.

Similarly, if you type [C-F]A, the software will select the points with the codes CA, DA, EA, FA, but not AA, BA, or GA.

'

This character facilitates point selection when codes contain one of the above operator characters. For example, we learned that typing *AB result in the selection of points with the codes like SAB or 3AB; but if your intention is to select the points with the code *AB, you must type '*AB.

For more complex selections, the explained characters can be used in combination with each other.

17- How to combine two or more point groups?

To combine two or more groups, start creating a new point group, type a name in the *Information* tab, go to the *Point Groups* tab, tick the checkboxes for the points you want to combine (Figure 44B), and press *OK*.

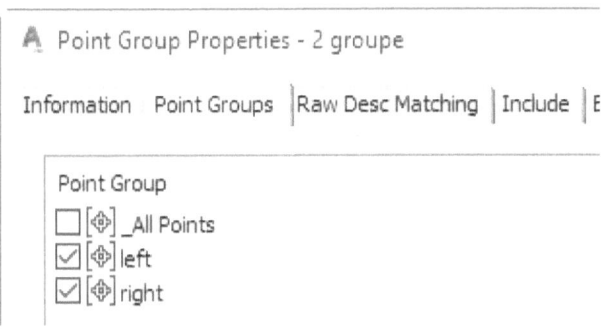

Figure B44

For practice, open the <u>combine point group</u> file from the folder named <u>project file</u>. This file contains two point groups named *Left* and *Right*, which you must combine into one group.

18- How identify the point group to which a point belongs?

To learn the point group of a specific point, select the point so that a tab named *COGO Point* followed by a point number appear in the ribbon. In the *Modify* panel of this tab, there is a button labeled *Point Group Properties*, which will show the point group to which the point belongs.

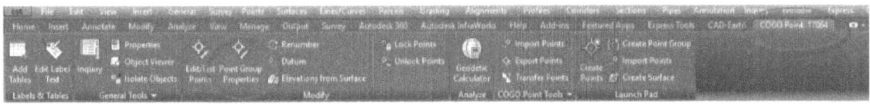

Figure B45

19- How to make the elevation label show a customized elevation if a certain condition is met?

For this exercise, open the file set elevation.txt from the folder points files. As you can see, some of the points in this files have an elevation of less than 2000 meters. Here, the goal is to make the elevation label add a 1000 meters to the displayed elevation if it is less than 2000 meters, all without altering the actual elevation values.

To do this, you need to define an *expression* with a conditional instruction.

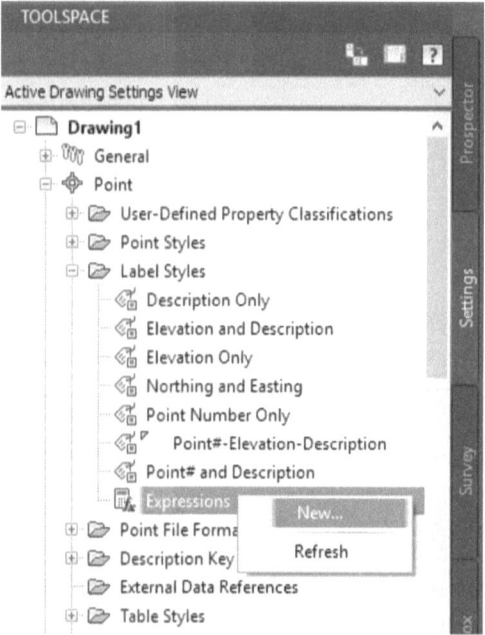

Figure B46

For this purpose, go the *TOOLSPACE* window, click on the *Settings* tab, expand the *Point* branch, and then the *Label Style* sub-branch. Right-click on *Expressions* and select *New...* .

This opens a window (Figure B47) where you can define the expression and its condition.

Figure B47

In the *New Expression* window, first, type a name for the expression to be defined, then go to the *Expression* field to define it. For this exercise, the expression should be in a conditional format and assert that if the elevation of a point is less than 2000, then it must be increased by 1000, otherwise it should remain unchanged. To create this expression, first click on the button No.1 (Figure B47) and select the function *IF*, then click on the button No.2 and insert *Point Elevation*. Next, type *<2000* in front of the inserted term to complete the condition and then type a comma to initiate the second part of the condition. The second part is where you define what happens if the condition is met, that is if the elevation of a point is less than 2000 meters. Here, you must again click on the button No.2 to select *Point Elevation* and then type *+1000*. When done, type another comma to initiate the third part of the condition, which determines what happens if the condition is not met. For this part, click on the button No.2 and select *Point Elevation* without adding anything else, which means the actual elevation of the point will be displayed. Finally, finish the expression with a parenthesis and press *OK*.

If done as instructed, the final expression should be:

IF (Point Elevation)<2000,{Point Elevation}+1000,{Point Elevation})

After creating the expression, go to the *Prospector* tab, right-click on the point group and select *Properties*. In the opened window (Figure B48), set the *Point Label Style* to *Elevation Only* and then select *Edit Current Selection* from the cascading menu to its right.

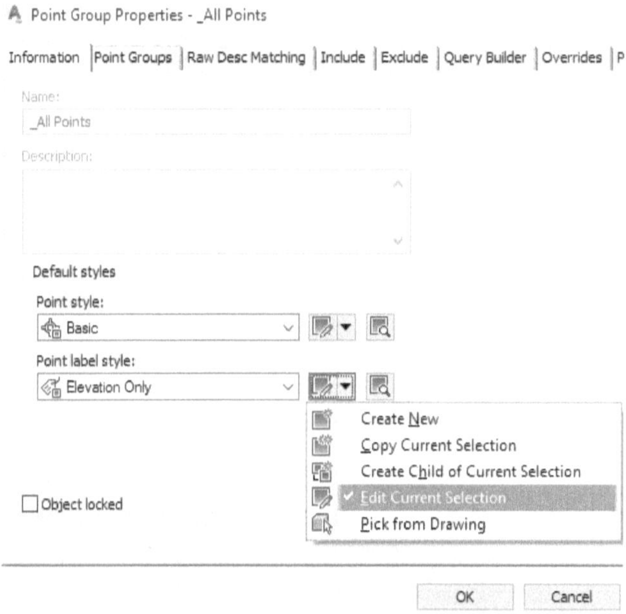

Figure B48

As shown in Figure B49, go the *Layout* tab and click on the text box in front of *Contents* to open the *Text Component Editor* window (Figure B50). In this window, first, delete the text written in the right pane and then browse the *Properties* menu for the name of the expression you created in the previous step. Click on the arrow button to move the selected expression to the right. Press *OK* to close all existing windows.

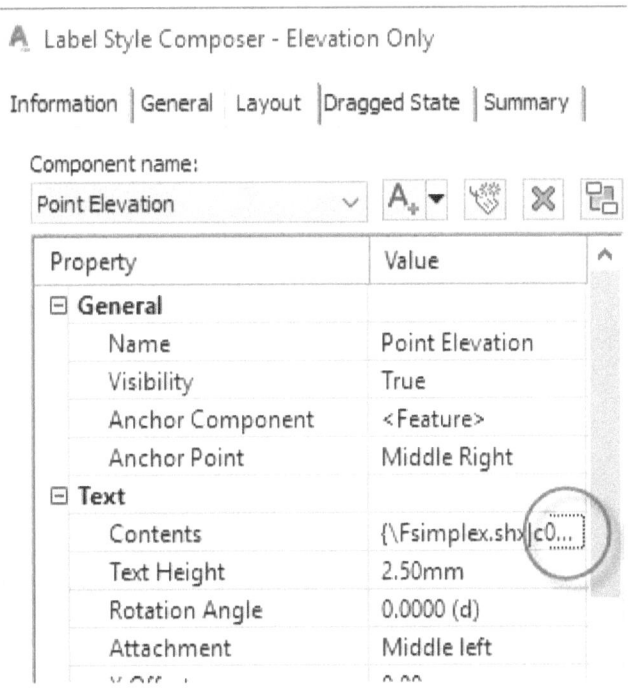

Figure B49

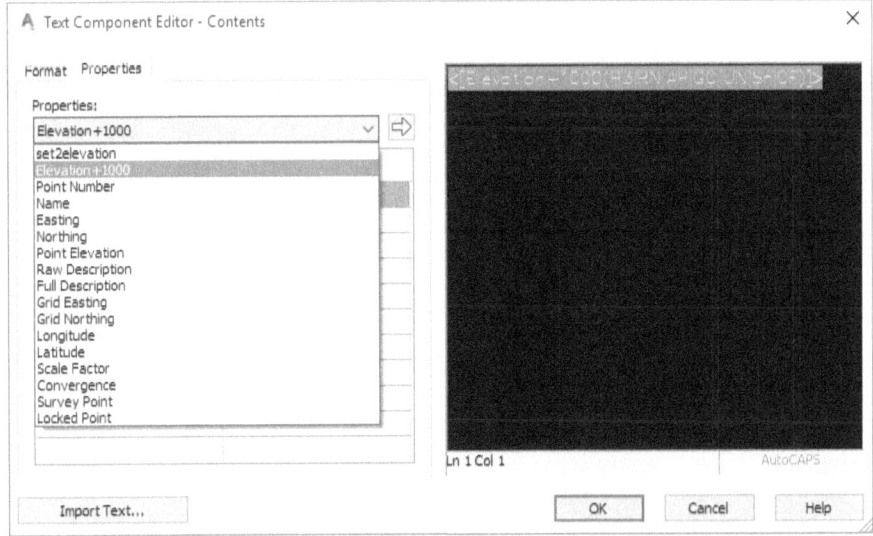

Figure B50

Now you can see the results in the labels. Figure B51, for example, shows that the label of the point 13745 displays an elevation of 2483.083, although its true elevation is 1483.083.

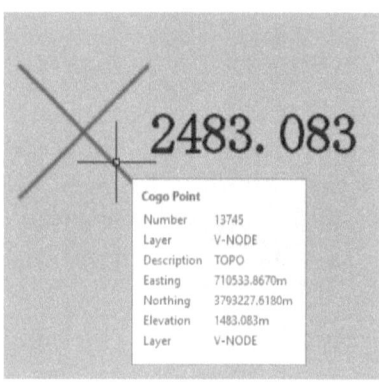

Figure B51

After learning other functions and some practice, you can apply multiple conditions of your choice to the point labels.

For example, you can type

={Point Elevation}+100

in the *expression* field to make the software increase the elevation label of all points by 100 meters, or type

IF(Easting>710500&Easting<710750,{Point   Elevation}+100,{Point Elevation}B00)

to make a 100-meter increase in the elevation label of all the points whose X coordinate is less than 710,500 and decrease the elevation label of other points by 100 meters.

20. How to create a point table displaying customized values instead of true values?

CIVIL3D allows you to create point tables with parameters such as UTM coordinates, geographic coordinates, elevation, etc. This can be done using the *Add Table* tool provided in the *Point* menu (Figure B52). Sometimes, it is convenient to change the information displayed in the table without actually making any change in the data. For example, suppose you want to create a point where all displayed elevations are 10 meters higher than their true values.

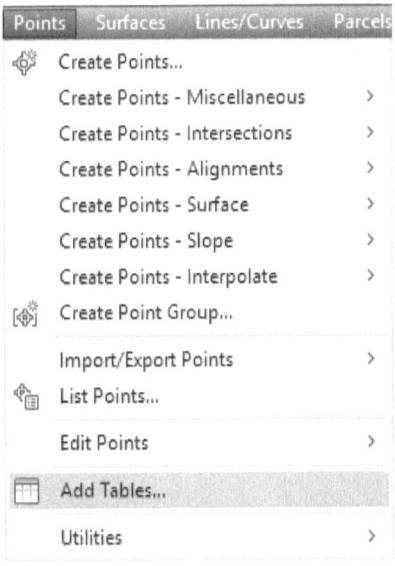

Figure B52

For this purpose, you have to create an *Expression* in the same way you did in the previous exercise (Figure B46), but this time use the expression {Point Elevation}+10.

For this exercise, open the file Point Table.dwg in the folder Project File. This file contains multiple points, for which we want to create a point table where all displayed elevations are raised by 10 meters.

After creating and saving the expression, open the *Points* menu and select *Add Table* (Figure B52). In the opened window (Figure B53), set the *Table Style* to the *PENZD format* and select *Edit Current Selection* from the cascading menu to the right.

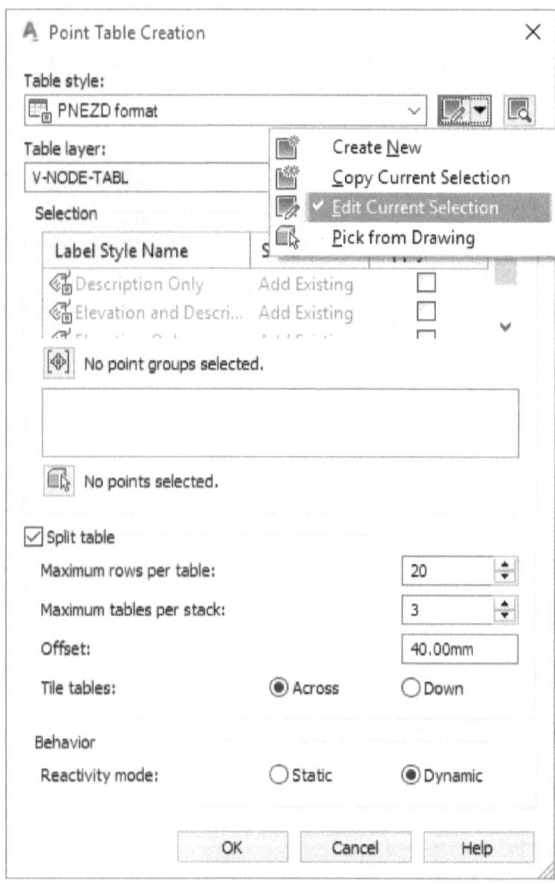

Figure B53

This opens the window shown in Figure B54, where you must go the *Data Properties* tab and find the *Structure* pane. In this pane, click on *Point Elevation* from the *Elevation* column to open the *Text Component Editor* window. In this window (Figure B55), first, delete the contents of the right pane and then select the previously created expression from the

Properties menu, and finally click on the arrow button to move the expression to the right. When done, press *OK*.

As shown in Figure B56, the elevation values displayed in the point table will be 10 meters higher than the actual elevations.

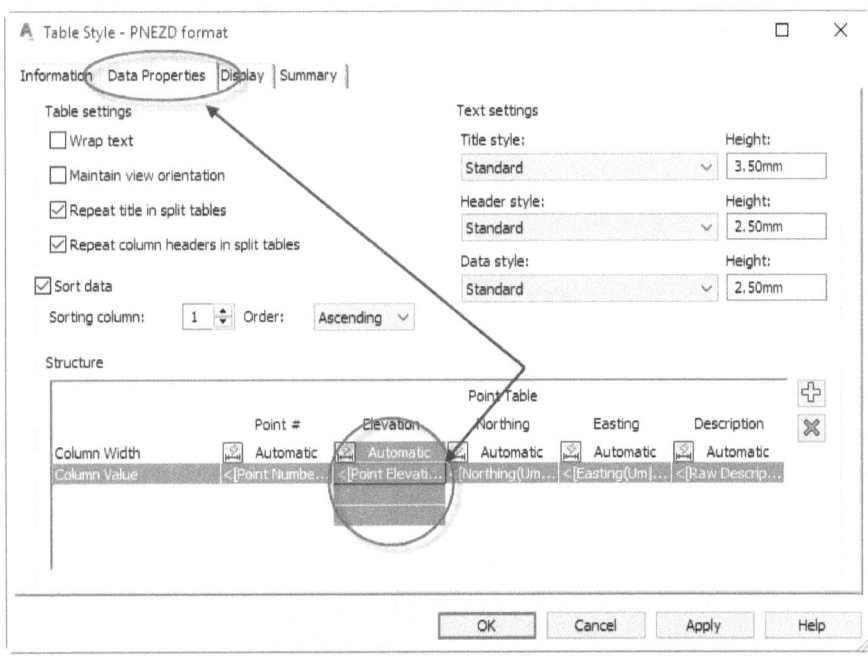

Figure B54

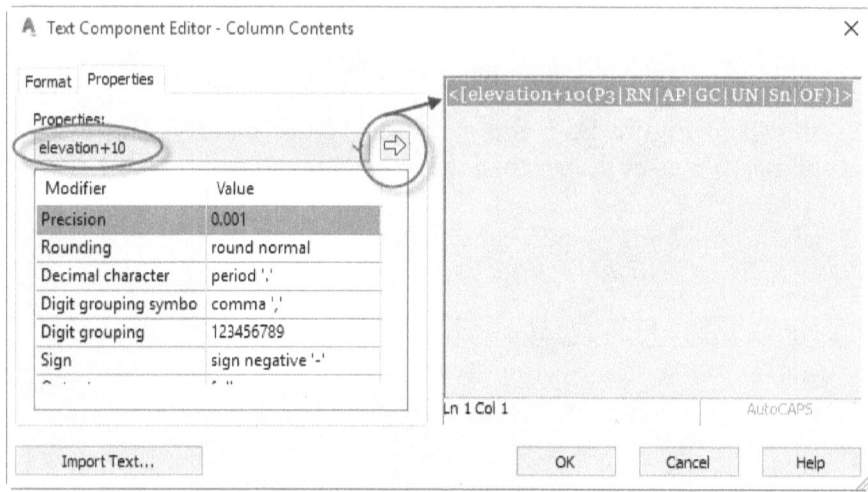

Figure B55

Note: in the *Point Table Creation* window, you can use the *Selection* box to specify the point group for which table will be drawn, or use the button below it to manually select the points.

9112	2467.539	3792997.81
9113	2456.299	3793101.80
9114	2457.379	3793073.84
9115	2456.339	3793040.34
9116	2461.989	3793003.34
9117	2453.499	3793105.89
9131	2477.899	3792942.14
9132	2473.069	3792969.73
9133	2482.369	3792963.47
13717	2489.645	3793103.04

Figure B56

21- Automatic marking of points

For this exercise, open the file <u>Point Marker.dwg</u> in the folder <u>Project File</u>. This file contains a number of points with the codes GAS, TIR, TREE, CHAH, GHANAT, and BM. Displayed in the corner of the drawing are the markings considered for each code (Figure B57).

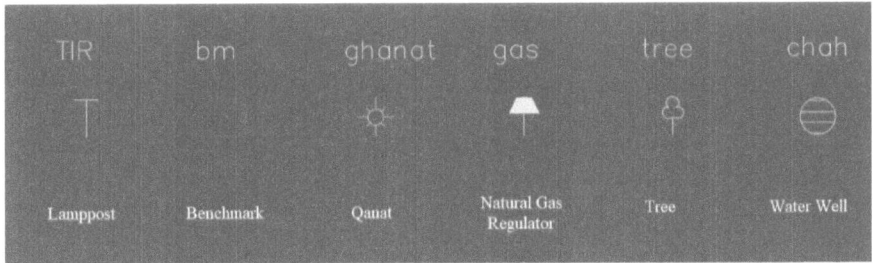

Figure B57

For automatic marking of features, you need to define a marker for each feature. The *Point Style* tool of the software provides a large group of blocks for defining markers, but those markers may not match the regional standards. AutoCAD allows you to define the marker of your choice through custom blocks. For this purpose, run the Block command in the command line (or use the shortcut B) to initiate building a block for markers. Executing this command opens the window shown in Figure B58.

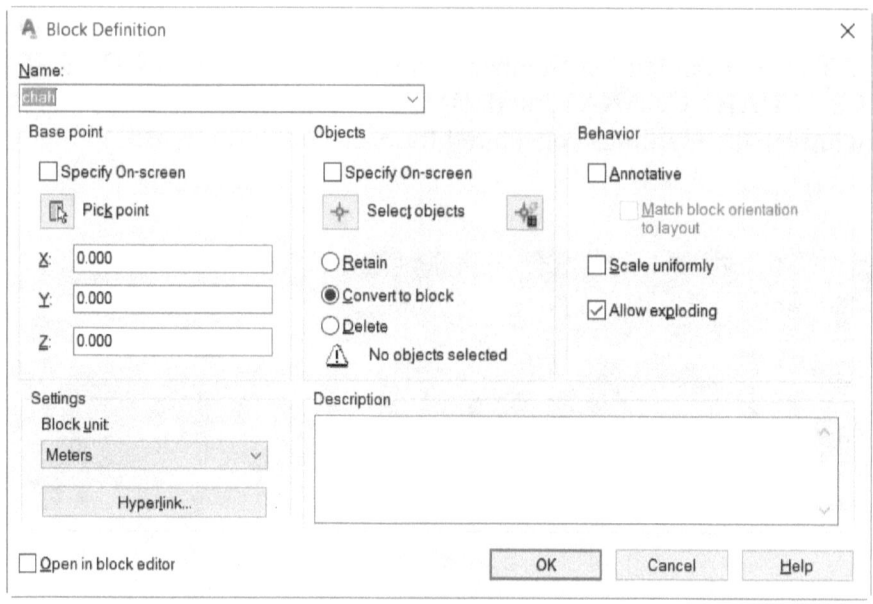

Figure B58

In this window, first, give your block a name, then press the *Select object* button to choose the components of the block, and finally press the *Pick point* button to specify the anchor point, i.e. the place where the block will be displayed. Once done, press *OK* to close the window.

Repeat this procedure to create one block for each feature (for convenience, name the blocks after the features).

The next step is to create a marker based on these blocks.

To create a marker, open the *TOOLSPACE* window, right-click on the point group of interest and select *Properties*. As shown in Figure B59, go the *Information* tab, click on the button next to *Point style* and select *Create New* from the opened cascading menu. In the *Information* tab of the opened window, type a name for the marker.

Figure B59

Then, as shown in Figure B60, go to the *Marker* tab, enable the option *Use AutoCAD BLOCK Symbol for the marker*, and then select the block that you created for the feature from the box below. Specify the size of the block in the *Size* field on the right, and finally press *OK* to close the window. Repeat the same process for each feature using the corresponding blocks (the ones you created in the previous step).

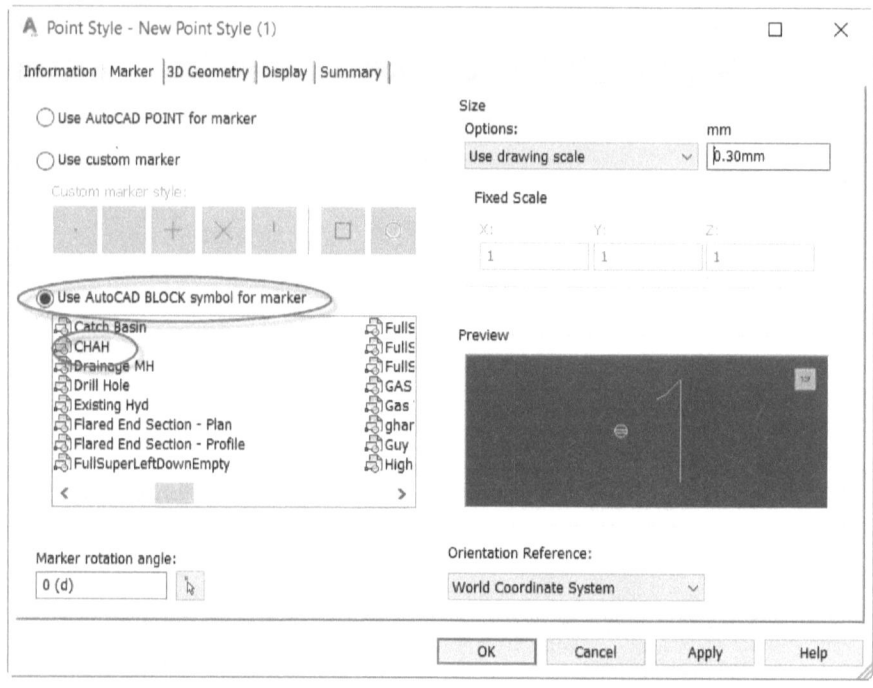

Figure B60

Next, open the *TOOLSPACE* window, go to the *Settings* tab (Figure B61), expand the *Point* branch, right-click on the *Description Key Sets*, and select *New*. In the opened window, enter a name for the new description key set and save it (it should appear under the *Description Key Sets* branch). Right-click on the set you just created and select *Edit Keys* (Figure B62).

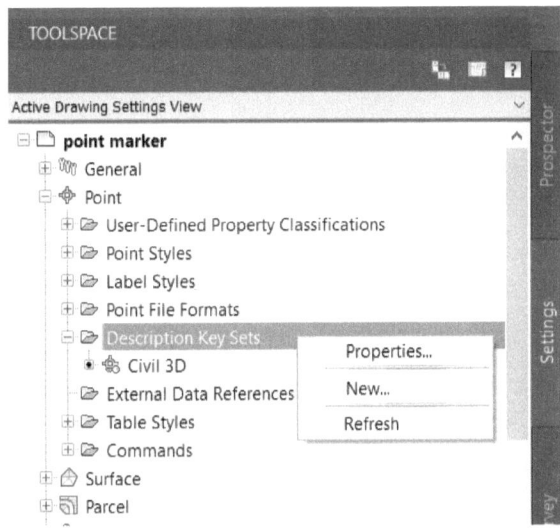

Figure B61

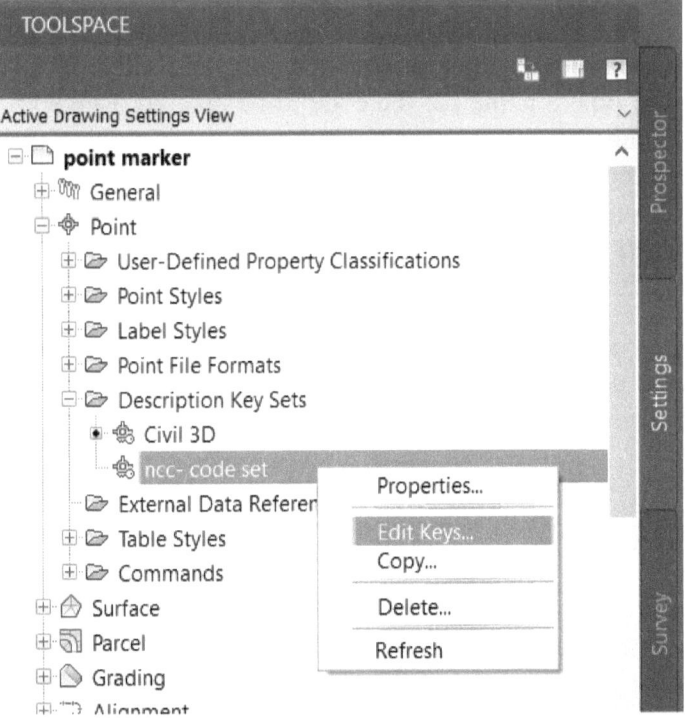

Figure B62

This opens a window for defining and editing codes, which is displayed in Figure B63.

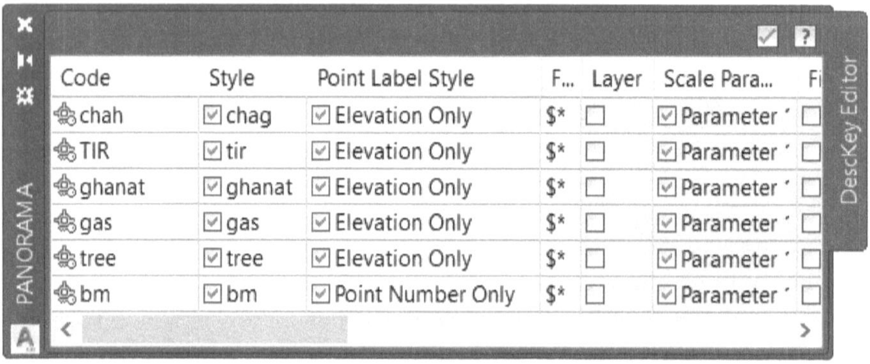

Figure B63

In this window, right click on the first code and select *New* to define a new code and type a name for the code to be created. Remember that the process of automatic marker attachment is case-sensitive, so the code defined here must be identical to the code of your points. In the *Style* tab, double-click on the style of each code and select the associated marker from the list. If needed, you can go to the *Point Label Style* tab and select a label for the marker. When done press *OK* to close the window.

Next, go back to the *Prospector* tab in the *TOOLSPACE* window, right-click on the desired point group, and select *Apply Description Keys* (Figure B64) to place the markers on the point.

If markers do not appear on the points, you may have made a mistake in code definition (most likely in using uppercase and lowercase letters).

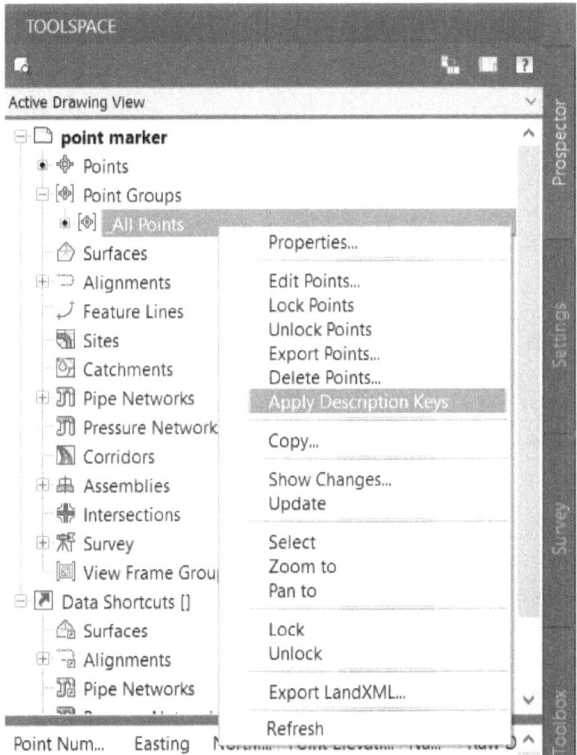

Figure B64

To change the size of the markers, right-click on the point group and select *Properties*, then open the *Point Style* menu and select *Edit* (Figure B65).

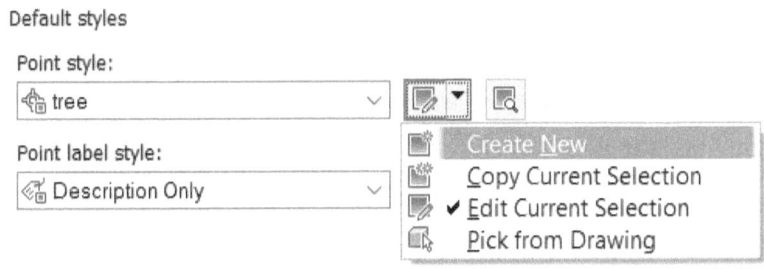

Figure B65

22- How to apply an elevation adjustment during export?

Occasionally, you may want to export the points of a drawing as a text file, but with a certain automatic adjustment made in the elevation (or other parameters) of your points. This requires creating an expression and then defining an export format accordingly.

Figure B66

In this exercise, we want to increase the elevation of points by 5 meters during export. To create the expression needed for this purpose, open the *Settings* tab in the *TOOLSPACE* window, expand the *Point* branch and then the *Label Style* sub-branch, right-click on *Expressions* and select *New...* (Figure B46). As shown in Figure B66, type a name for the expression and then create {Point Elevation}+5 in the *Expression* box the way you learned in previous exercises. You now have created an Expression that increases the elevation by 5 meters.

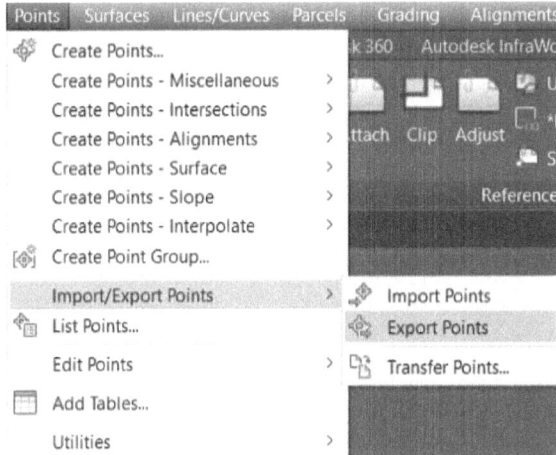

Figure B67

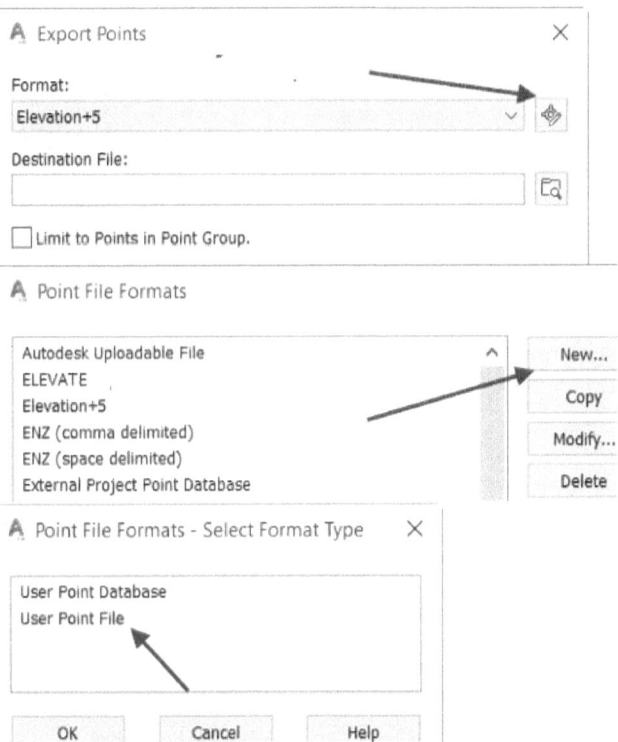

Figure B68

Now, open the *Points* menu and select *Import/Export Points* and then *Export Points* (Figure B67). Click on the buttons marked in Figure B68 to open the *Point File Format* window shown in Figure B69.

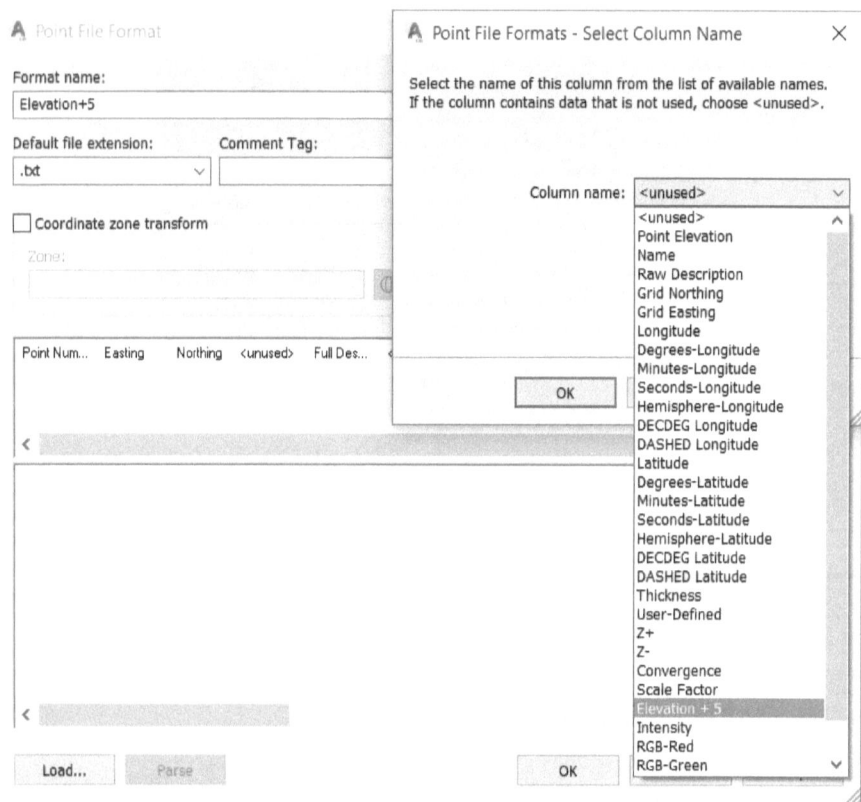

Figure B69

In this window, first type a name in the *Format name* field then set the first column below to *Point Number*, the second column to *Easting*, the third column to *Northing*, the fourth column to the expression you just created (here *Elevation+5*), and the fifth column to *Full Description*. Save this format so that it can be used for export.

You can also edit the typical PENZD format and replace the *Elevation* in the fourth column with your custom-made expression.

23- How to prune the points based on proximity in position and elevation

Suppose that a drawing consists of numerous duplicate or near duplicate points created by mistake or for any reason. These points have no importance in terms of measurement or design and must be pruned to facilitate drawing and avoid error. For this purpose, these points must first be detected and then removed. In the software, these steps are separated to prevent the automatic removal of those points that the software identifies as duplicate but the user views as the start or a part of a breakline. Although the software provides some tools to enhance the user's selection power, only the detection is automatic and the actual removal must be carried out by the user. However, the software handles the most difficult part of this process, which is the detection of duplicate points.

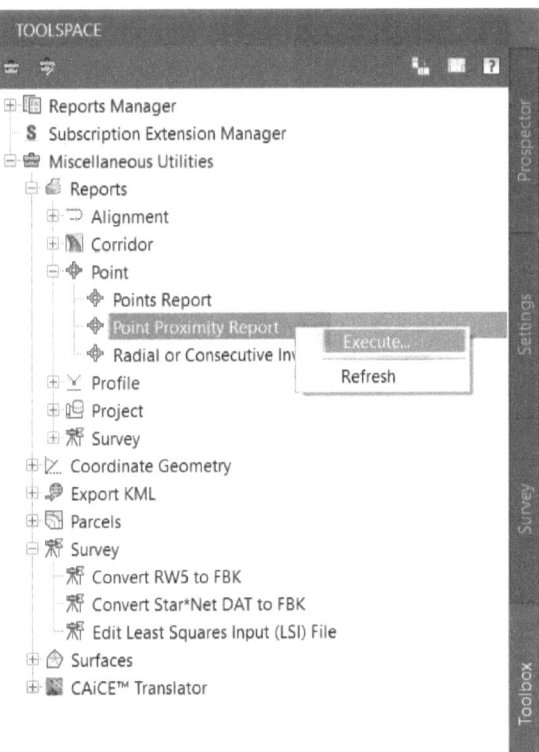

Figure B70

For this exercise, open the file <u>Clear point proximity.dwg</u> from the folder <u>Project file</u>. As you can see, this file contains 25622 points. Here, the goal is to remove one of any pair of points that are less than 50cm apart (horizontally) and have less than 5cm elevation difference. Although this can be done manually, the process will be painstakingly time-consuming. The more convenient alternative is to use the *Point Proximity Reporter* tool. For this purpose, open the *TOOLSPACE* window and then the *Toolbox* tab. Under the *Miscellaneous Utilities* branch, expand the *Reports* and then *Points* sub-branches. Right-click on *Point Proximity Report* and select *Execute* (Figure B70),

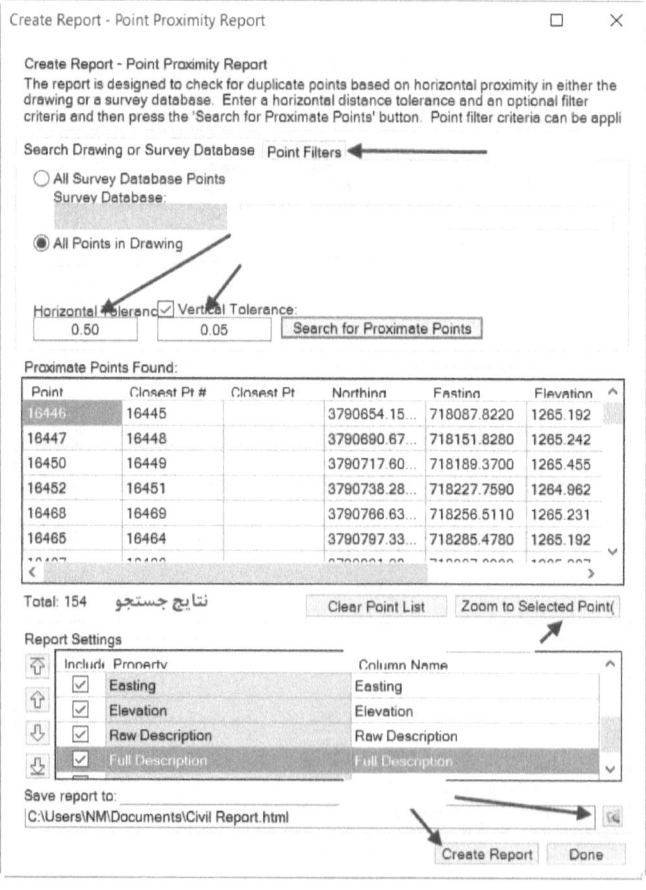

Figure B71

This opens the window shown in Figure (B71). In this window, first activate the *All Points in Drawing* radio button, then type 0.50 in the horizontal tolerance field and 0.05 in the vertical tolerance field, and finally click on the *Search for Proximate Points* button. At the end of this search, the software shows all the points that meet the defined conditions, i.e. those with a horizontal distance of less than 50 cm and vertical distance of less than 5cm from each other. In the file of this exercise, 154 such points must be found. In the search results pane, the first column provides the number of the first point and the second column shows the number of the point that meets the defined proximity conditions. You can select any of these points and click on the *Zoom to Selected Point* button to check its condition. You can also use the *Create Report* bottom at the bottom of the window to export the search results as a text or Excel file.

This search may not be enough to meet your needs. For example, the drawing may have two points that are closely-positioned but are both needed as one is related to a well and the other represents its adjoining wall. To resolve this problem, AutoCAD provides some tools to enhance the user's control over the search process. At the top of this window, there is a tab named *Point Filter*, where you can filter your search results. As shown in Figure B72, you can tick the *Raw description* box and type the point code of your choice (in this example, *TOPO*) and then click on the *Apply Point Filter* button to limit the search results to the points with the specified code. In this exercise, filtering the results for the *TOPO* code reduces the number of duplicates from 154 to 94.

You can also use the *Point Name* and *Elevation Range* options to limit the search results to specific ranges of point numbers or elevation.

If you have drawn a boundary by a polyline, you can activate the *Area* option and select that polyline to limit the search results to the corresponding boundary.

Create Report - Point Proximity Report   ☐   ×

Create Report - Point Proximity Report
The report is designed to check for duplicate points based on horizontal proximity in either the
drawing or a survey database. Enter a horizontal distance tolerance and an optional filter
criteria and then press the 'Search for Proximate Points' button. Point filter criteria can be appli

Search Drawing or Survey Database   Point Filters

☐ Point Name   ☐ Area:

                                          Select Figure / Parcel / Closed Polyline A

☑ Raw Description(s):   ☐ User-Defined Property Classification:
TOPO

                    From: To: Property Name: Property Value(s):
☐ Elevation Range

                                               Apply Point Filters

Proximate Points Found:

Point	Closest Pt #	Closest Pt	Northing	Easting	Elevation	^
16836	16837		3791263.05...	718180.5510	1265.995	
16709	16705		3791746.62...	718536.2450	1264.328	
14378	14379		3791864.45...	718351.8010	1268.694	
14291	14290		3791872.42...	718128.6660	1277.118	
14288	14287		3791950.77...	718147.8990	1277.409	
14407	14406		3792107.88...	718264.2310	1268.730	v
2014	2010		3792010.00	719101.0000	1110.007	

Total: 94   Clear Point List   Zoom to Selected Point(

Report Settings

⇪	Include	Property	Column Name	^
⇧	☑	Easting	Easting	
	☑	Elevation	Elevation	
⇩	☑	Raw Description	Raw Description	
⇟	☑	Full Description	Full Description	v

Save report to:
C:\Users\NM\Documents\Civil Report.html   ☜

                                        Create Report   Done

Figure B72

Once done with browsing, checking, and adjusting the search results
with filters, you can delete the duplicate points from the original file using
excel or in any way you see fit.

24- How to produce length and angle outputs for setting out with theodolite?

In the projects where points cannot be set out by coordinates (because reference benchmarks are somehow lost, drawing is not georeferenced, there is no access to a total station, etc.), this must operation must be carried out classically using length and angle values.

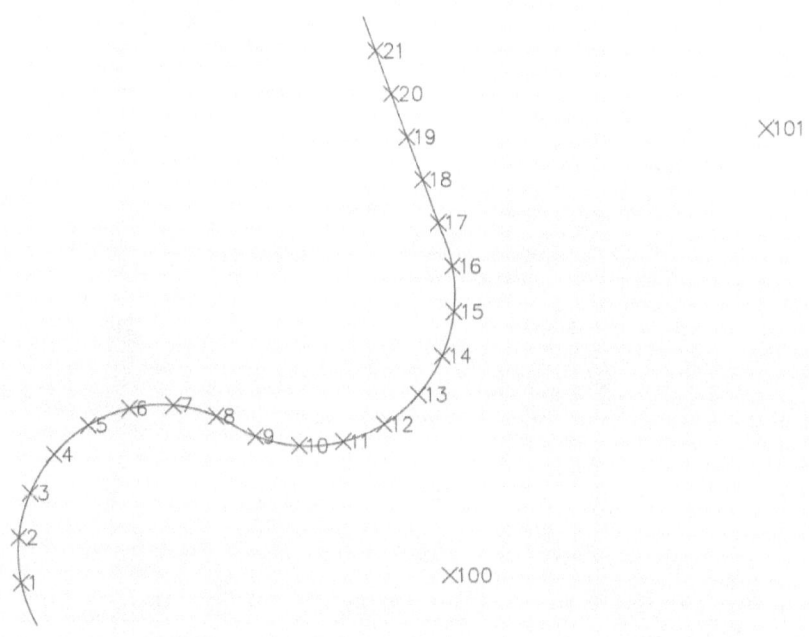

Figure B73

For this exercise, suppose that in a simple road construction project with the plot shown in Figure B73, the goal is to set out the road axis by deploying the theodolite on the Point100 and zeroing on the Point101. This requires a list of lengths and angles for setting out points 1 to 21 from the said station.

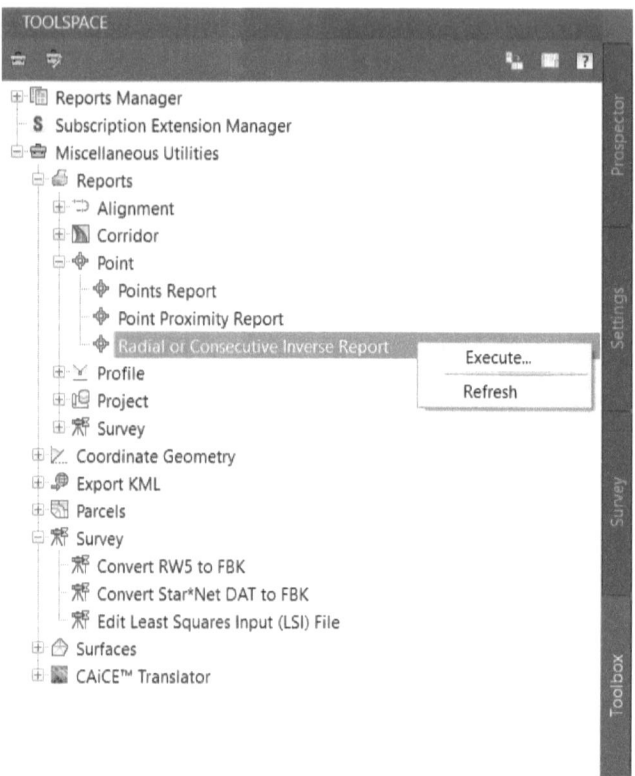

Figure B74

Open the file point-axe-LH.dwg from the folder Project File. To extract the longitudinal and angular coordinates of target points, open the *TOOLSPACE* window and the *Toolbox* tab, (Figure B74). Under the *Miscellaneous Utilities* branch, expand the *Reports* and then *Points* sub-branches. Right-click on *Radial or Consecutive Inverse Report* and select *Execute* to open the window shown in Figure B75.

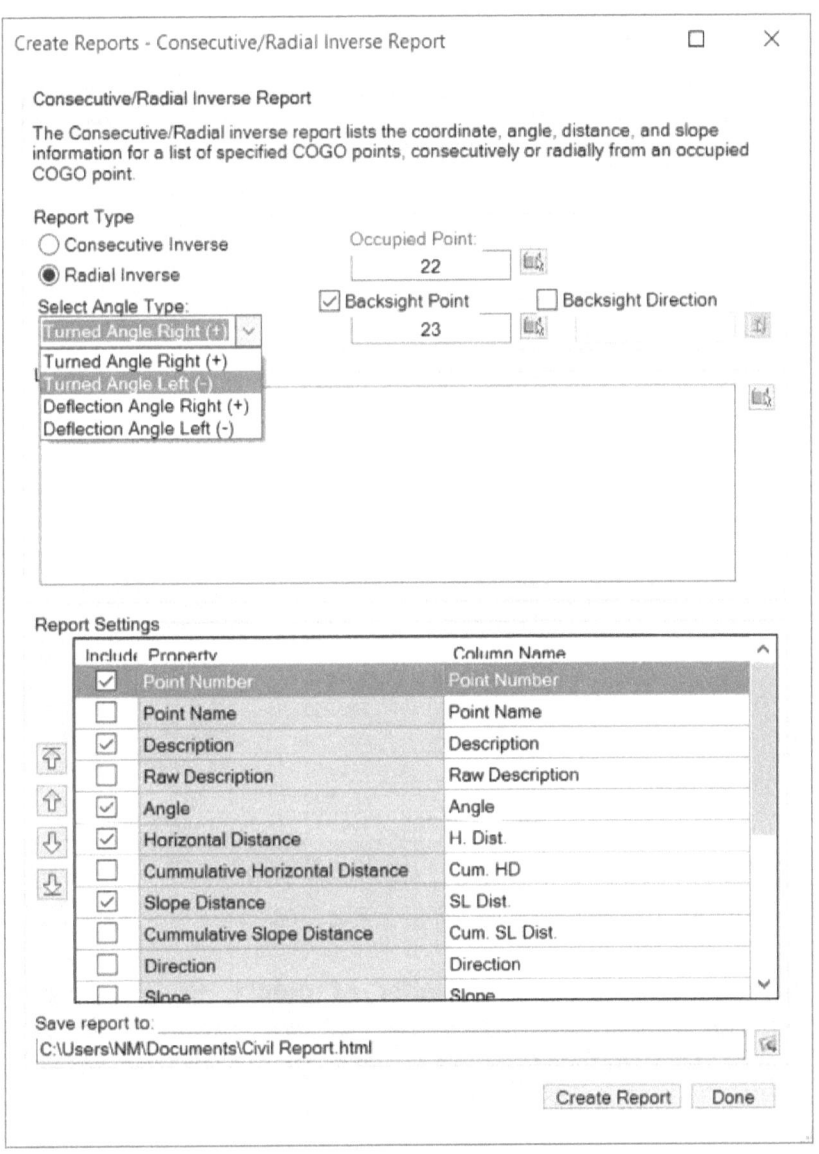

Figure B75

In this window, click on the icon in front of *Occupied Point* to specify the current station, use the *Backsight Point* option to specify the back sight point, and use the *Select Angle Type* to specify the angle direction.

Set the parameters needed in the output file (e.g. horizontal distance and angle, point number and code, etc.) in the *Report Settings* box.

Finally, press the *Create Report* button to generate the output file. In the end, the result should be as shown in Figure B76.

Point Number	Angle	H. Dist.	SL Dist.
1	233.0308 (d)	385.571	385.571
2	238.9032 (d)	389.109	389.109
3	244.7658 (d)	384.338	384.338
4	250.4830 (d)	371.417	371.417
5	255.8942 (d)	350.779	350.779
6	260.7819 (d)	323.162	323.162
7	264.8185 (d)	289.655	289.655
8	267.4691 (d)	251.825	251.825
9	268.7855 (d)	212.24	212.24
10	273.7831 (d)	176.135	176.135
11	284.5711 (d)	150.643	150.643
12	299.9039 (d)	144.28	144.28
13	313.7672 (d)	159.892	159.892
14	321.9105 (d)	191.064	191.064
15	324.8614 (d)	229.424	229.424
16	324.4019 (d)	269.221	269.221
17	321.9925 (d)	307.257	307.257
18	319.8756 (d)	345.403	345.403
19	318.1817 (d)	383.927	383.927
20	316.7976 (d)	422.725	422.725
21	315.6468 (d)	461.729	461.729

Figure B76

25- How to determine the number of points, the maximum and minimum elevations, the mean slope, and two- and three-dimensional area of a created surface?

For this exercise, open the file named Contour.dwg in the folder Project File. As you can see, this file contains the drawing of a surface. To access the abovementioned information about the surface, open the *TOOLSPACE* window and the *Prospector* tab, right-click on the name of the surface, and select *Surface Properties* (Figure B77) to open the *Properties* window (Figure B78).

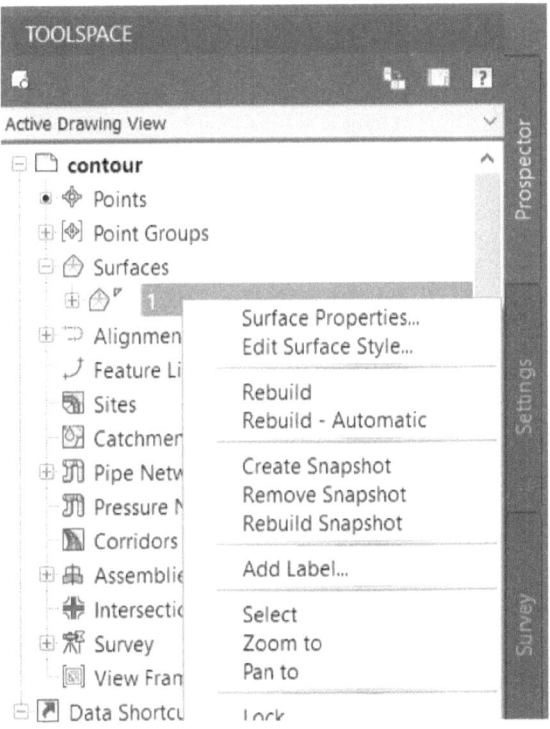

Figure B77

In the *Statistics* tab of this window, you can find a summary of important properties, including the number of points, the minimum and maximum X and Y values of the area, the minimum and maximum elevations, two and three dimensional surface area, the minimum slope, maximum slope, and mean slope of the area, the number of triangles, the minimum and maximum triangle area, and the minimum and maximum triangle length.

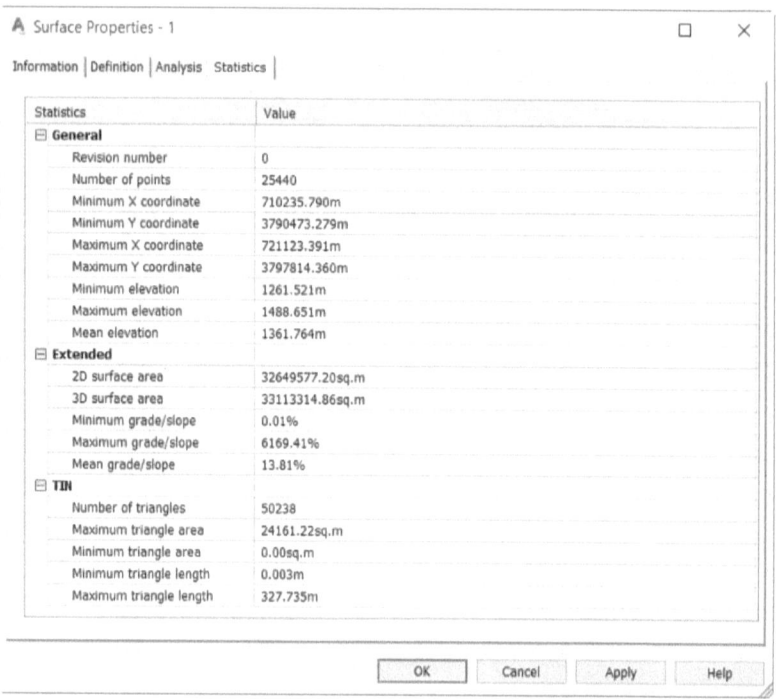

Figure B78

26- How to smooth the drawn contour lines?

To smooth the created contour lines, go the *Prospector* tab from the *TOOLSPACE* window, right-click on the surface of interest and select *Edit Surface Style* to open the window shown in Figure C1.

In this window, open to the *Contours* tab, expand the *Contour Smoothing* branch, set the *Smooth Contours* option to *True*, and then adjust the degree of smoothing with the slider on the bottom.

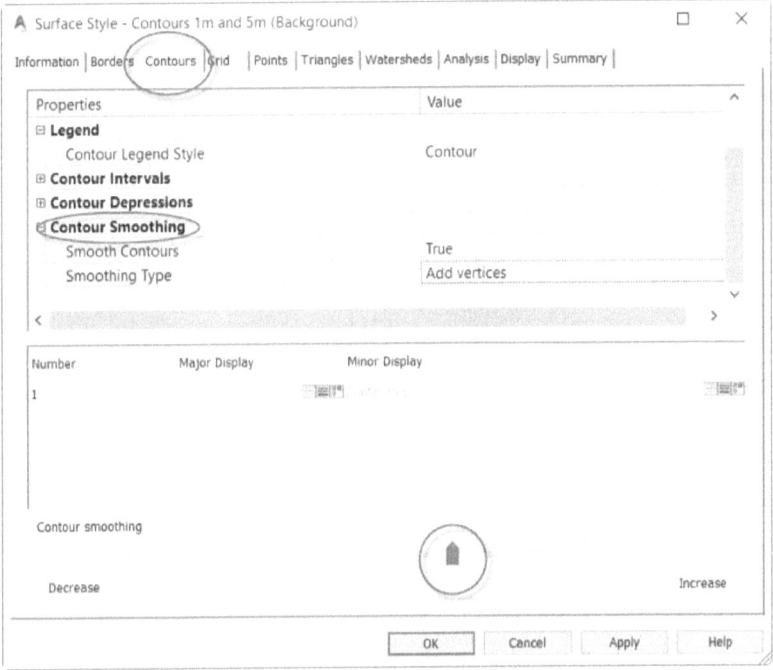

Figure C1

27- How to calculate the volume of mass between two elevation levels without creating a cross section or intermediate surface?

One of the standard ways to highlight the elevation variations of an area in a drawing is to use colors. The most typical approach in this regard is to divide the distance between the highlight and lowest elevations to several ranges, and portray the areas belonging to each range with a different shade of a color. Note that it is customary to use lighter shades for lower elevations and darker shades for higher elevations.

To perform this coloration in the software, open the *TOOLSPACE* window and the *Prospector* tab, right-click on the surface and select *Edit Surface Style*, and then go to the *Display* tab and turn off all layers except *Elevation* (Figure C2).

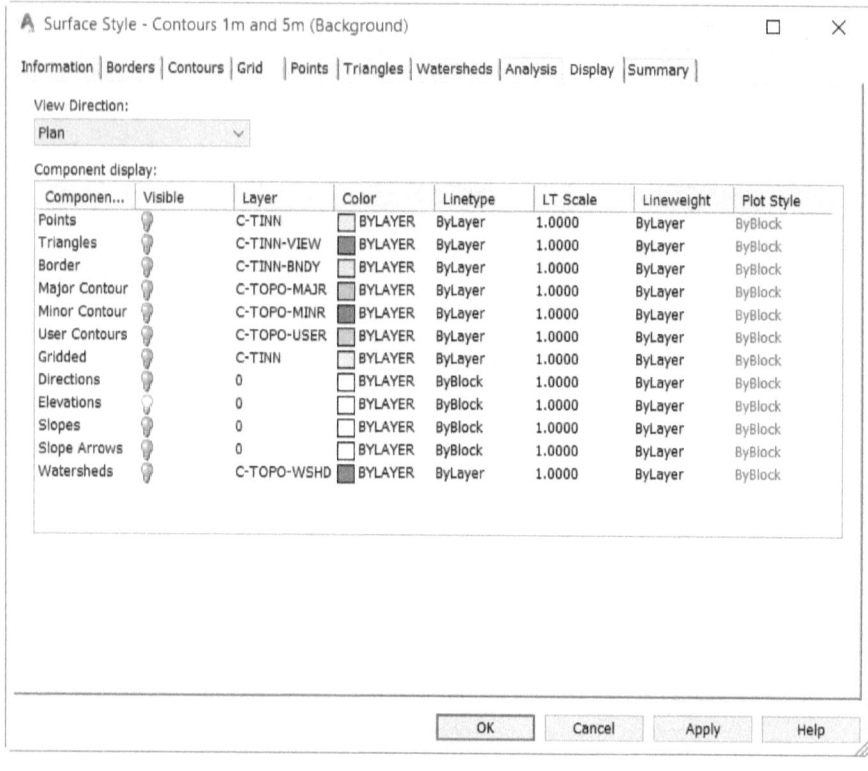

Figure C2

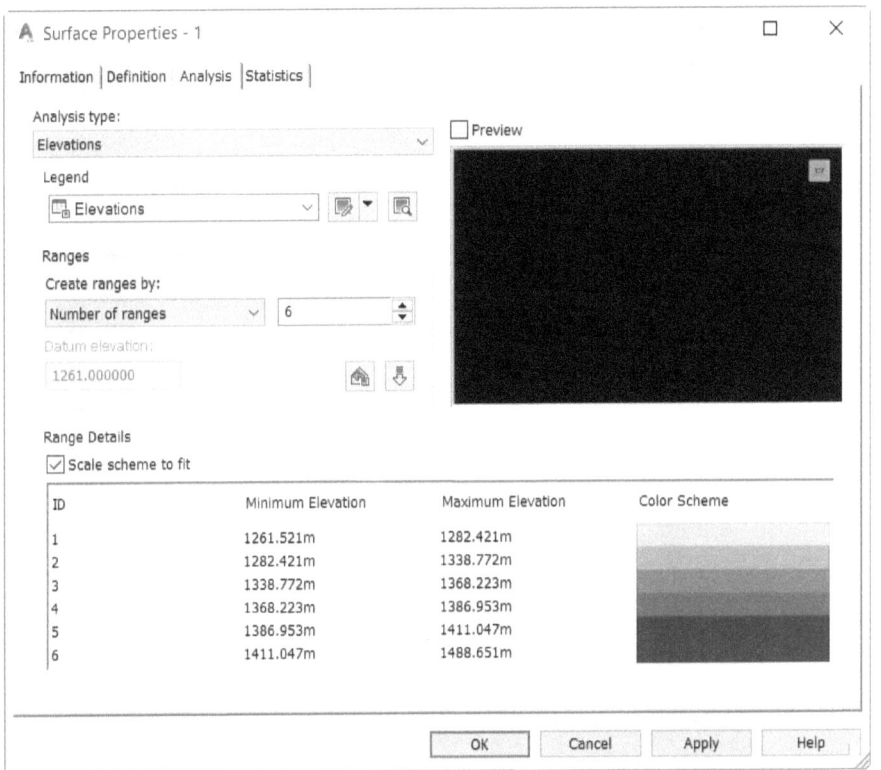

Figure C3

Next, return to the *TOOLSPACE* window and the *Prospector* tab, right-click on the surface and select *Surface Properties* to open the window shown in Figure C3. In this window, go to the *Analysis* tab and set the *Analysis type* to *Elevations*. To divide the elevation into a specific number of ranges, set the *Create ranges by* menu to *Number of ranges*, and type the desired number of ranges in the field to the right. Then, click on the arrow button to add the selection to the box below. This box shows the highest and lowest elevation levels to be colored and the color scheme to be used for this purpose and allow you to make adjustments as needed. Once finished with the settings, press *OK* to apply the changes.

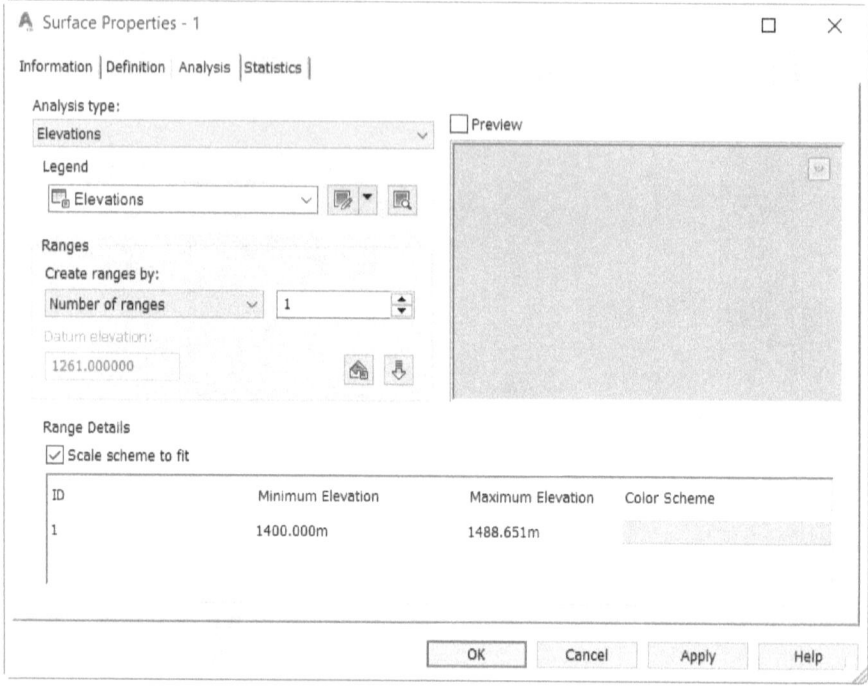

Figure C4

If you want the ranges to be of a certain size, set the *Create ranges by* option to *Ranges Interval* and type the desired size in the field.

One of the most important applications of the described procedure is to calculate the volume of mass between two elevation levels. For this exercise, open the file Contour.dwg from the folder Project File. In this excavation project, the goal is to reduce the maximum elevation from 1488.65 to 1400.00. To estimate the amount of excavation needed to achieve this goal, open *Surface Properties* window as described above, set the *Create ranges by* menu to *Number of Ranges*, type 1 in the right field, press the arrow button, set the *Minimum Elevation* to 1400.00, and press *OK*. Doing this will highlight all areas with an elevation of between 1400.00 and 1488.651 with the selected color (here green) as shown in Figure C5.

To determine the area and volume of this colored zone, you can simply insert a legend table containing the area and volume information of the zone.

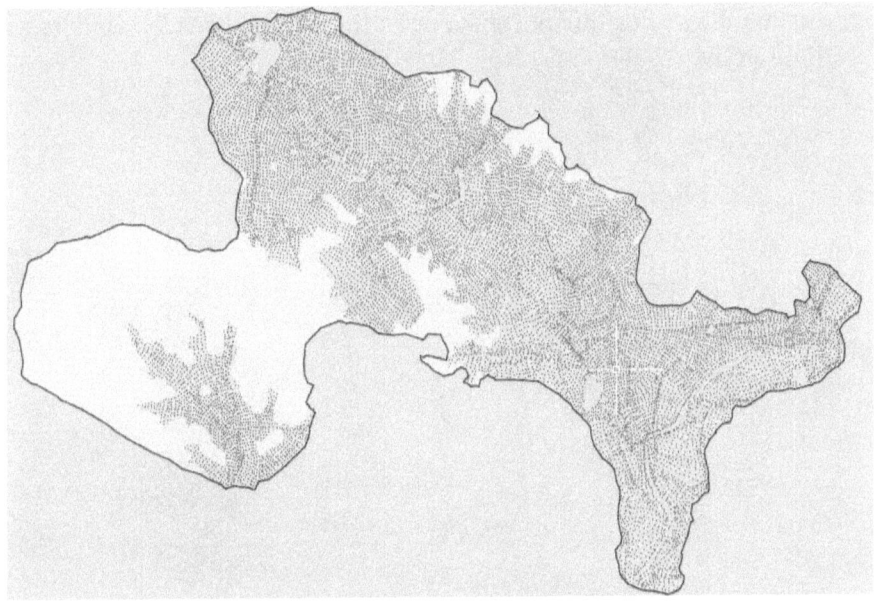

Figure C5

To insert this table, open the *Surface* menu, click on *Add Legend Table...* and select *Elevations* (Figure C5-6). After selecting the *Static* or *Dynamic* mode, click on a point on the drawing to insert the table.

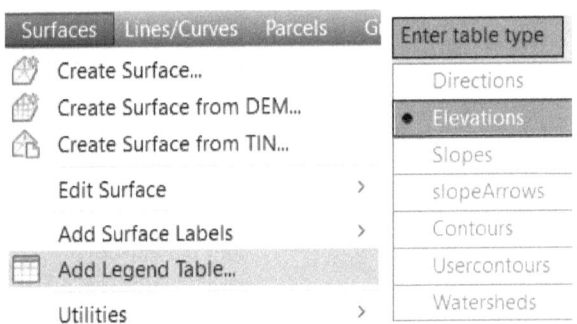

Figure C5-6

But this table will only show the relevant area. Inserting a column to display the volume information requires some changes in the Surface

Properties window. For this purpose, open the TOOLSPACE window and right-click on the cr

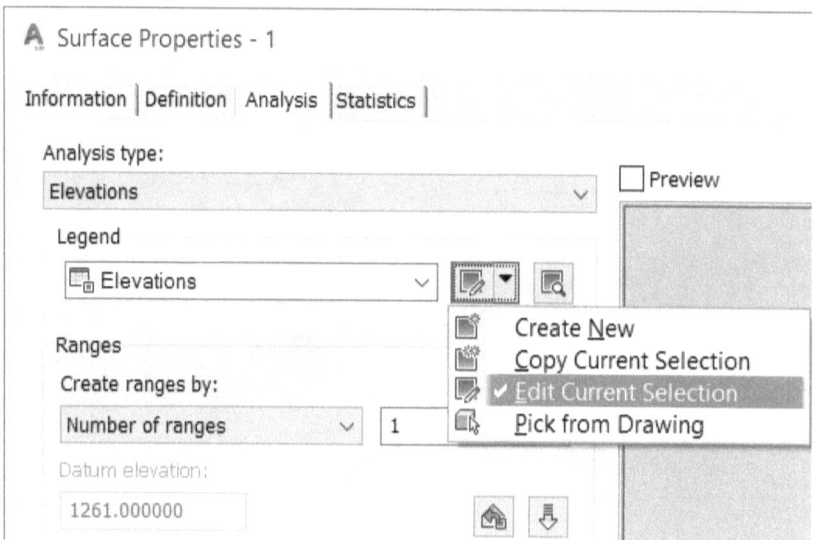

Figure C6

eated zone and select Surface Properties. In the opened window, click on the cascading menu to the right of Elevations, and select Edit Current Selection (Figure C6) to open the Table Style window (Figure C7). In this window, go to the Data *Properties* tab and click on the + icon in the *Structure* box to add another column to the table.

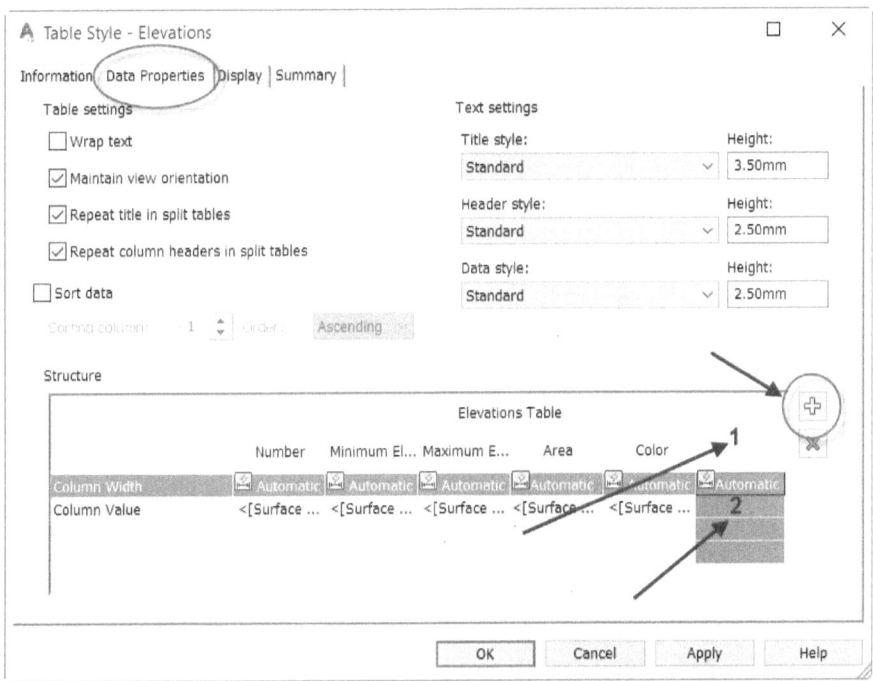

Figure C7

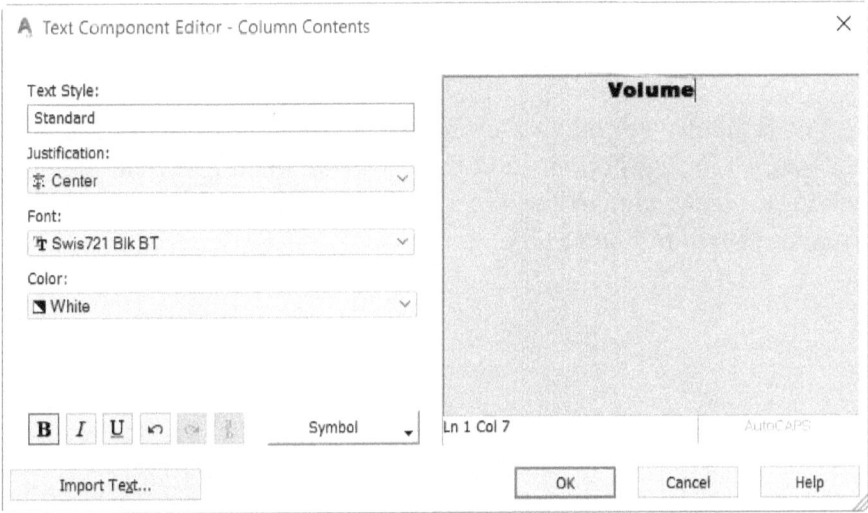

Figure C8

Clicking on the said button adds a new column to the right of the last column (which should be *Color* by default). To give the added column a name, double-click on the button No.1 in Figure C7 to open the window shown in Figure C8, type *Volume* in the right pane and press *OK*.

Then double-click on the button No.2 in Figure C7 to open the window shown in Figure C9.

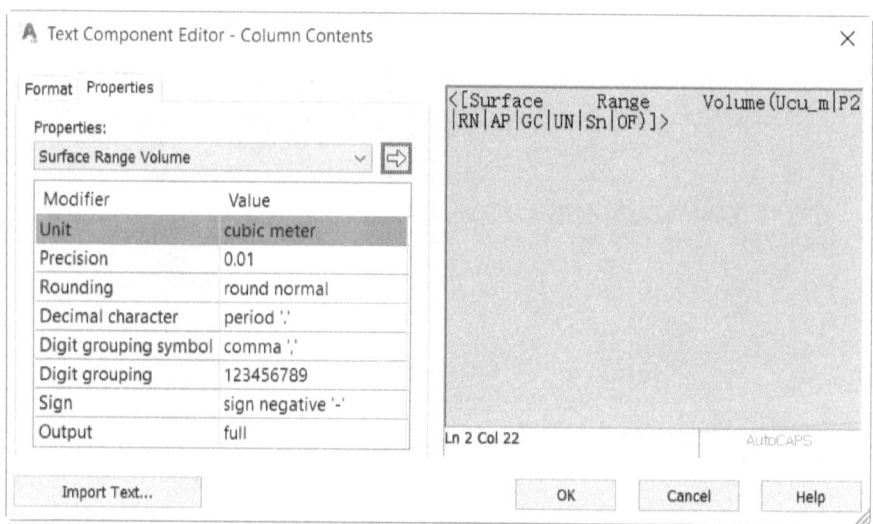

Figure C9

In this window, go the *Properties* tab, open the *Properties* menu and select *Surface Range Volume*, and then click on the arrow icon to add the selection to the right. After pressing OK, table information will be updated as shown in Figure C10.

Elevations Table					
Number	Minimum Elevation	Maximum Elevation	Area	Color	**Volume**
1	1400.00	1488.65	8951260.66		187692966.21

Figure C10

The *Volume* column of this table now displays the volume of mass between the elevation levels 1400 and 1488.651 (should be 187692966.21 m³), and its *Area* column shows the area of this zone.

28- How to create a point by interpolating elevation from a surface?

Often, surveyors need to create large numbers of points by interpolating elevations from available surfaces. For this purpose, after creating the surface and contour lines, you must open the *Points* menu, click on *Create Points - Surface* and then select *Random Points* (Figure C11). Now, if you click on the point of interest, the elevation of that point will be calculated by interpolation.

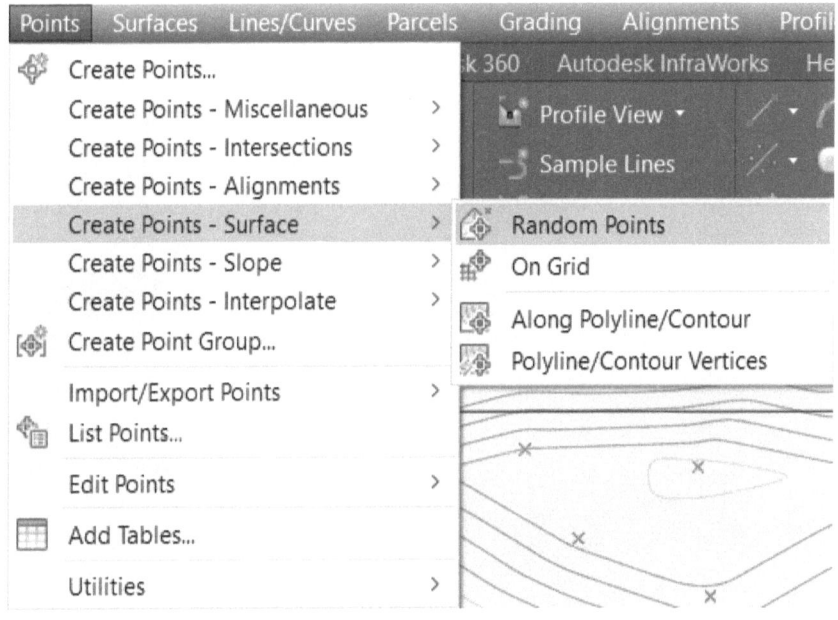

Figure C11

29- How to convert the data of a topographic survey into a grid map?

One of the common problems in surveying projects is that while the ultimate goal is to obtain a grid map (40×40 for example), terrain, features or other obstacles force the surveyor to proceed with a topographic approach and then convert the collected data into a grid. For this process to succeed, the coordinates of grid vertices should be known and the surface made with the converted points should match the surface made with the topographic points.

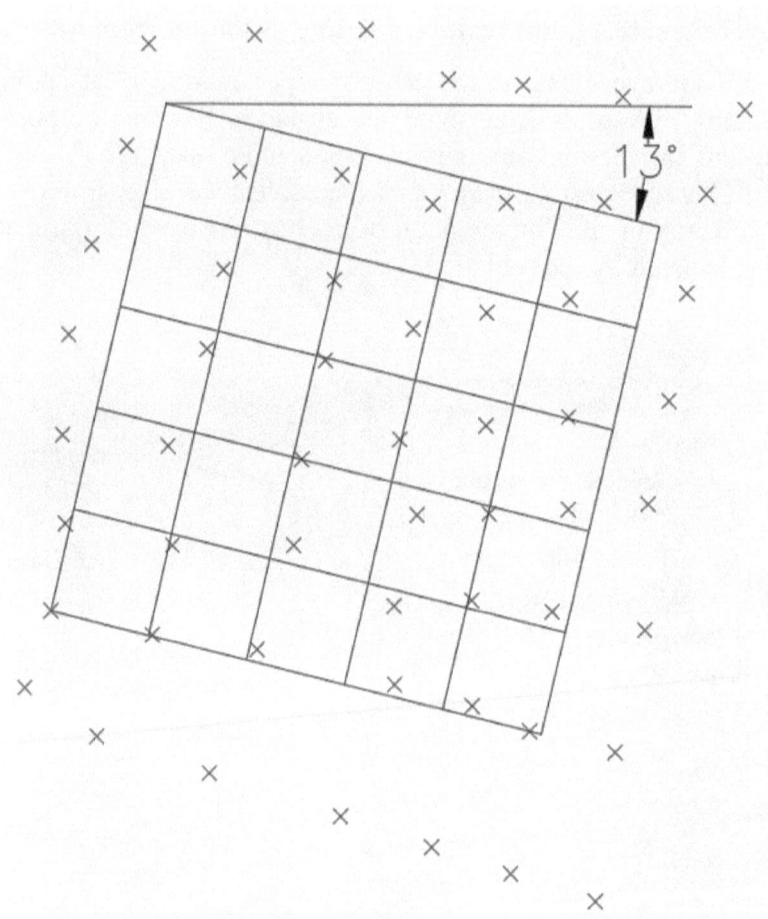

Figure C12

For this exercise, open the file topo to grid.dwg from the folder Project File. As shown in Figure C12, this file contains a 40×40m grid, which has 13 degrees rotation with respect to the X-axis. This file also contains the information of the points that have been surveyed topographically in an effort to obtain this grid, but are actually quite off the grid nodes. To convert this topographic survey to a grid map, you have to derive the elevation of each grid node from the surveyed surface. Hence, you need to first create the surface of the surveyed points, and then derive the elevations of grid nodes from this surface. To do so, open the *Points* menu and click on *Create Points - Surface* and then select *On Grid*

(Figure C13). This opens a message on the command line asking from you to *Specify a grid basepoint*.

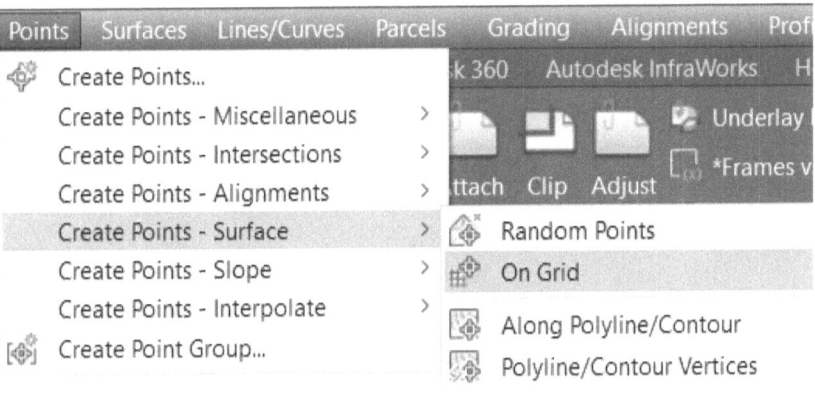

Figure C13

In response to this message, you must click on the point from which you want the grid to start. In this exercise, click on the southwest point shown in the figure.

Then, another message will ask you to specify the grid's rotation angle. In this exercise, this angle is -13, because the grid has 13 degrees rotation in clockwise direction.

Grid rotation <0.0000 (d)

This angle can also be specified by clicking on the three vertices of the rotation angle. The next step is to response to the following messages, which ask you to specify the grid spacing along X and Y directions. In this exercise, these spacing values are both 40 meters.

Grid X spacing <40.0000>: 40

Grid Y spacing <40.0000>: 40

The next message will ask you to specify the end point of the grid. Since we chose the southwest corner as the start point, we have to specify the northeast corner as the end point.

The software then ask that whether you want to change the rotation angle or grid spacing. Type Y to edit these parameters, or N to confirm your inputs and go to the next step. The next step is to create points according to the definitions provided earlier. Here, the software will display a message asking you to *Enter a point description*. Type a name to be used as description and press Enter to start creating points.

At this stage, it is recommended to give the points a new code so that they can be easily assigned to a point group or exported as a text file (as was discussed in previous exercises).

30- How to calculate the surface area of a slope?

The easiest way to calculate the surface area of a slope is to use the surface tool. In projects where the area of a slope is more important than the two-dimensional area (e.g. when calculating the quantity of the concrete cast or the material used on a slope), you just have to survey the slope, create a surface based on its points, and view the surface area on the *Surface Properties* window. For this exercise, open the file named shirvani.dwg from the folder Project File. This file contains the points surveyed on the corners of a slope with aim of calculating its area. With a horizontal dimension of 20m×30m, this slope has a two-dimensional area of 600m^2, but, naturally, the surface area will be greater than this figure. To determine the difference, first create a surface based on those points. Then, open the *TOOLSPACE* window and the *Prospector* tab, right-click on the created surface and select *Surface Properties* to open the *Surface Properties* window (Figure C14). In this window, go to the *Statistics* tab, where you can find the two-dimensional and three-dimensional area by expanding the *Extended* branch. As you can see, the three-dimensional area of this particular slope is 27.85m^2 more than its two-dimensional area.

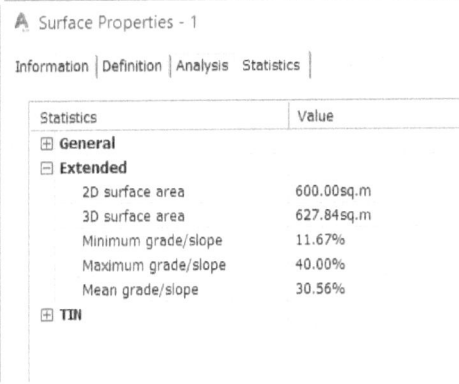

Figure C14

31- How to create a final as-built plot based on initial and secondary surveys (without conducting a new survey)

In many linear development projects, such as gas pipeline, water pipeline, and road construction, it is common to perform an initial survey over the project path (such as the one illustrated in Figure C15), and later perform a secondary survey to record the exact details of the work performed and create as-built plots. It is essential for as-built plots to cover not only the project site but also the surrounding area. During this process, surveyors need to combine the initial and secondary surfaces in a way that the points of the initial surface that are also present on the secondary surface are removed from the final plots. This process results in a new surface such as the one shown in Figure C15.

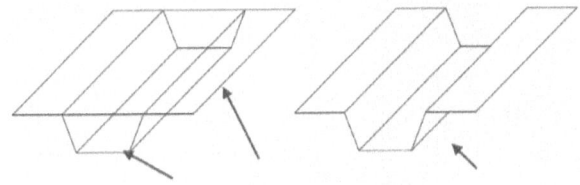

Figure C15

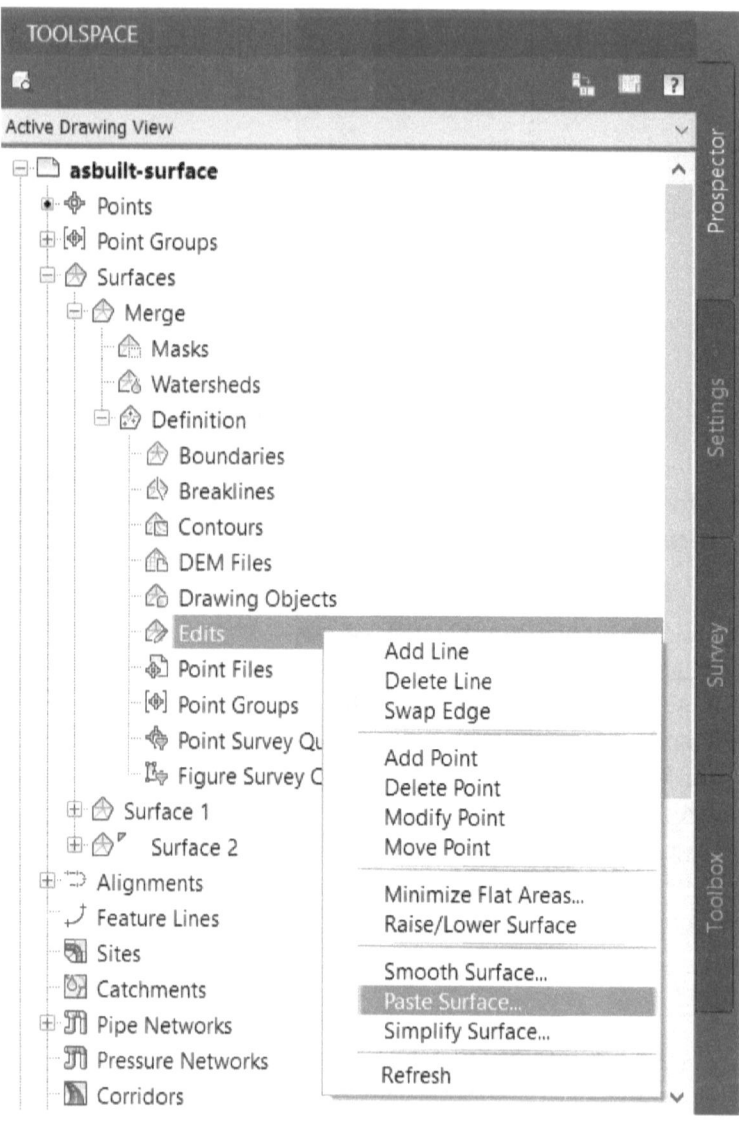

Figure C16

For this exercise, open the file <u>asbuilt-surface.dwg</u> from the folder <u>Project File</u>. This file contains an initial survey and a secondary survey of an area, which are named *Surface 1* and *Surface 2* respectively. The goal is to create a final as-build plot based on these two surfaces. For this purpose, first create another surface with an arbitrary name. Then, go the

TOOLSPACE window and open the *Prospector* tab, and select the surface you just created. Expand the branches shown in Figure C16. After expanding the *Definition* branch, right-click on *Edits*, and select *Paste Surface*. This opens a window, where you must first select *Surface 1* and press enter, then repeat the procedure and select *Surface 2*. By doing so, you make the software remove the points of the second surface from the first one, thus creating an as-built plot as shown in Figure C17.

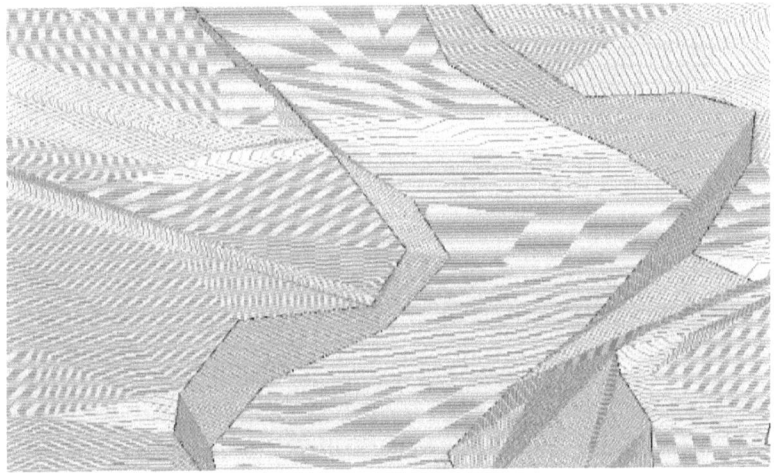

Figure C17

For better visibility, it is better to turn off *Surface 1* and *Surface 2* so that only the new surface is displayed. With this method, you no longer need to perform another survey for as-built plots or even remove the shared points manually. As will be described in later exercises, you can also create cross sectional plots of this surface and export them as a text file.

32- How to make Civil3D automatically draw the best possible boundary of a surface with least distance from points?

One of the common complaints of Civil 3D users is that, if you import a series of points, create a surface, and introduce the points to the created surface, Civil 3D will not automatically draw the boundary by crossing a line through the extreme points (as in Figure C18). Thus, user has to draw the boundary manually with polylines and introduce them to the surface. Indeed, this process can become very time consuming, especially if the project involves a great number of points.

Figure C18

This problem can be simply resolved by defining the maximum length of triangulation.

To do this, when in the *TOOLSPACE* window, right-click on the surface of interest and select *Surface Properties*. Go to the *Definition* tab and expand the *Build* branch in the *Definition Options* box (Figure C19). Then, set the *Use maximum triangle length* option to *Yes*. There, there is a field in front of *Maximum triangle length*, where you must type the maximum triangulation length that fits your project. The recommended maximum length for a 1:2000 map for example is 40. Finally press OK to close the window.

You can use the file boundary.dwg in the folder Project File to exercise this procedure.

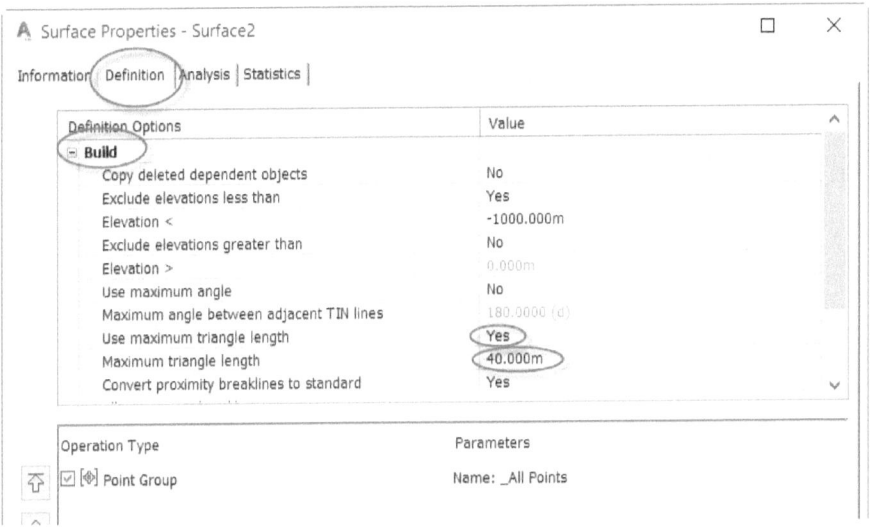

Figure C19

33- How to make elevation labels display customized values?

Sometimes it is necessary to make elevation labels display customized values with slight differences from the real values. For this exercise, open the file C-lable.dwg from the folder Project File. This file contains the contour lines of a region. Here, the goal is to make the elevation labels display the elevations 10 meters more than what they really are in the drawing. For this purpose, you should first create an *Expression* for representing elevation plus 10 meters.

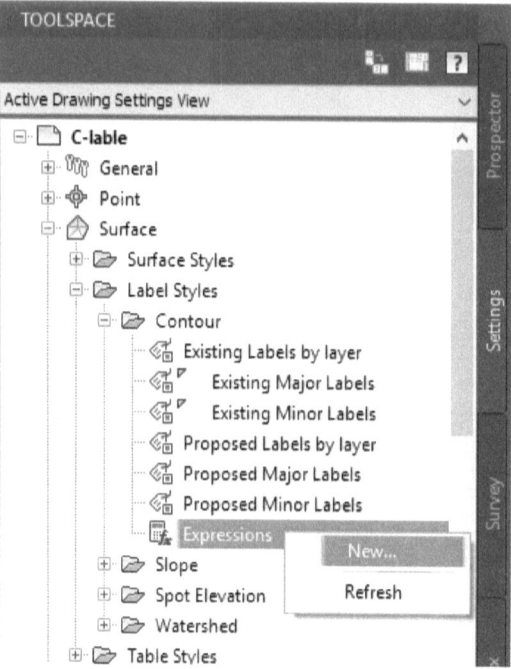

Figure C20

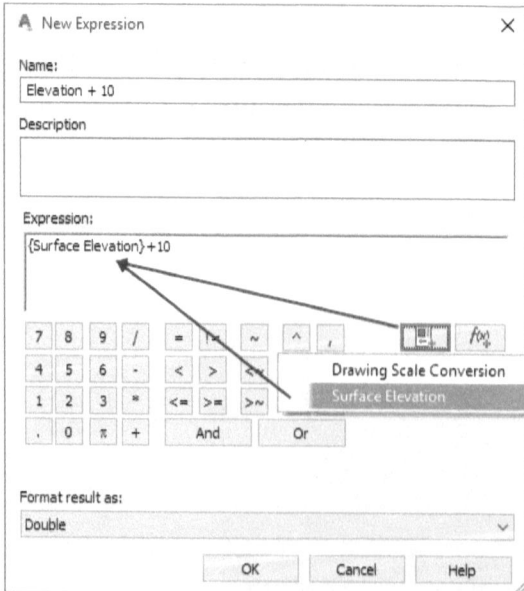

Figure C21

For this purpose, open to the *TOOLSPACE* window and the *Settings* tab, expand the *Surface* branch and then *Label Styles* and *Contour* sub-branches, right-click on *Expressions*, and select New (Figure C20). This will open the *New Expression* window (Figure C21). In this window, first type a name for the expression to be created, then use the button shown in Figure C21 to add a *Surface Elevation* expression to the *Expression box*. Next, type +10 in front of the added expression and press OK. After creating this expression, go to the *Prospector* tab in the *TOOLSPACE* window, expand the *Surfaces* branch, right-click on the surface of interest, and select *Add Label* (Figure C22). This opens the window shown in Figure C23. In this window, set the *Feature* menu to *Surface* and the *Label Type* menu to *Contour - Single* or *Contour Multiple*. Next, open the cascading menu next to the menu titled *Major contour label style* and select *Create New* (see Figure C23). This opens another window, which is displayed in Figure C24.

In this window, first type a name in the *Information* tab, then go to the *Layout* tab, and click on the field next to the box called *Contents* under the *Text* branch. This opens the window shown in Figure C25. In this window, delete the text written in the right pane, select the created expression from the *Properties* menu, and click on the arrow button to move it to the right.

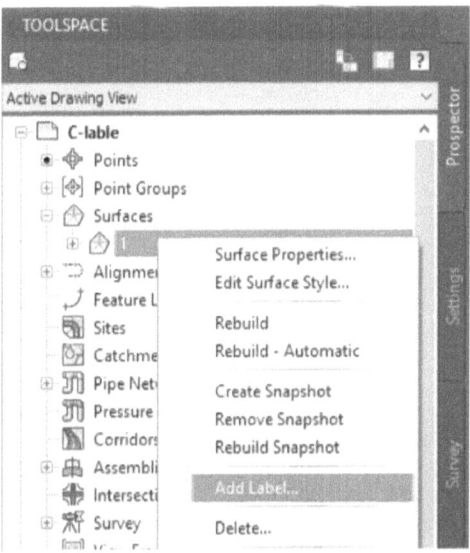

Figure C22

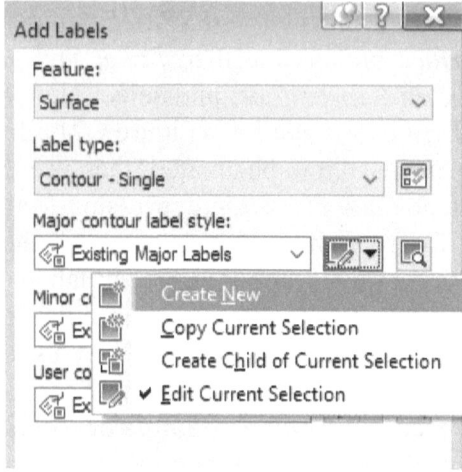

Figure C23

Close the windows in the order that they were opened, until reaching the *Add Label* window. Repeat the described process for *Minor Contour* and use the *Add* button to insert the label.

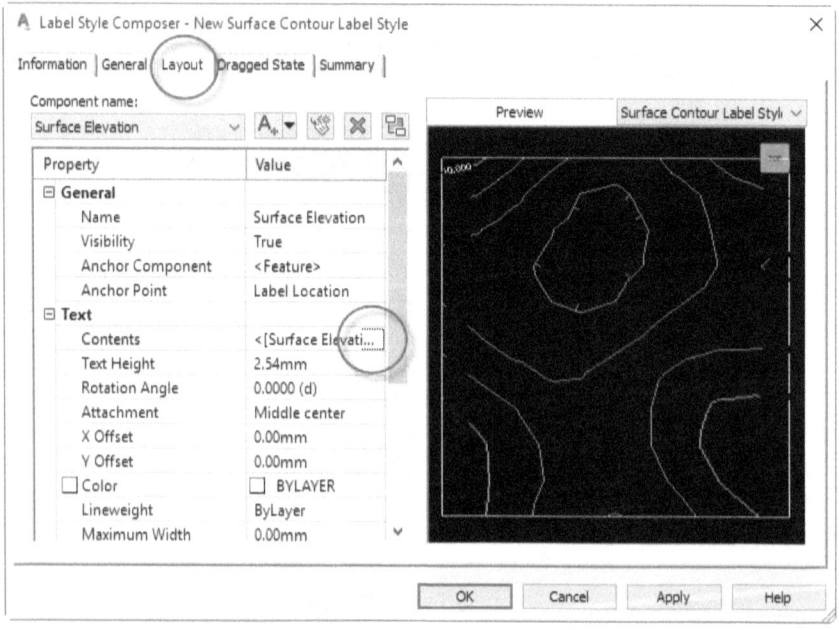

Figure C24

You can also create a conditional *Expression* (as is explained in previous exercises) for summing the elevations of less than a specified threshold with a certain value and summing others with another value.

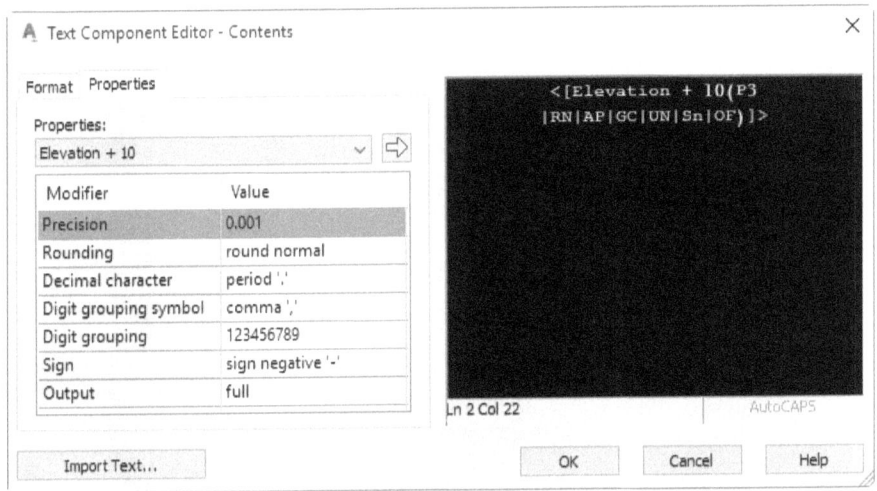

Figure C25

34- How to create a surface from a contour map in which contour lines are in polyline form?

For this exercise, open the file polyline-contour.dwg from the folder Project File. As you can see, this file contains a series of contour lines, from which we want to derive outputs such as transverse and longitudinal profiles. The first step for any such task is to create a surface based on the available polylines. Note that in this process, polylines should have an elevation. As you know, each contour represents an elevation that applies to all points on that contour, so each polyline must also have an elevation, and that elevation must match the corresponding contour line.

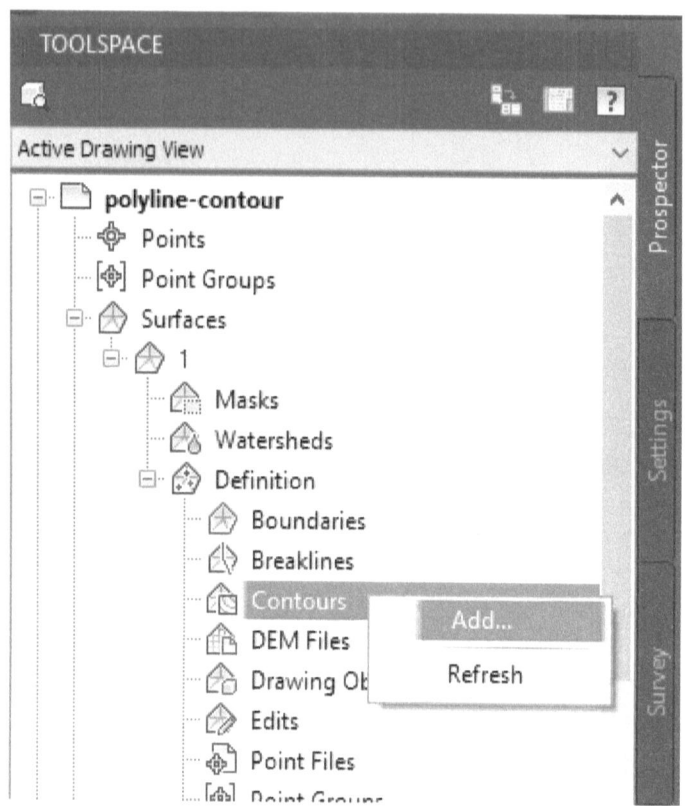

Figure C26

After opening the file of this exercise in CIVIL3D, create a surface and then open the *Definition* branch through the path shown in Figure C26, right-click on *Contours*, click on *Add* and then select all polylines. After defining the boundary of the surface, right-click on the created surface and select *Edit Surface Style*. From there, go the Display tab and turn on the *Major Contour* and *Minor Contour* layers (Figure C27). You can also turn off the polyline layer by using the *Layoff* command and selecting one of the major and minor polylines. In the *Summary* tab of the same window, you can adjust *Contour Intervals* to reduce or increase the spacing between contour lines.

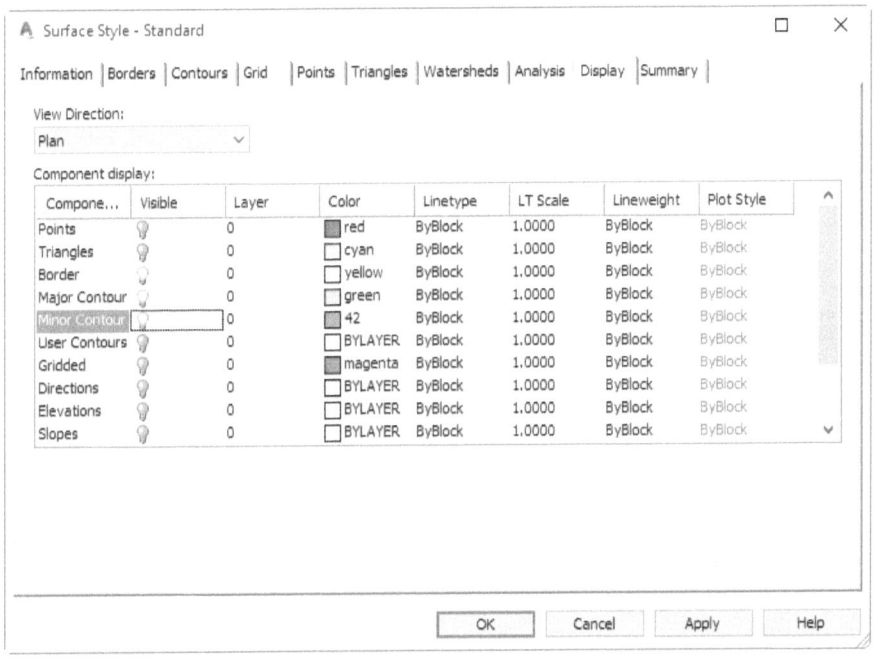

Figure C27

35- How to identify the surface zones with a slope of more than 35% ?

In many projects, designers and developers have to identify the area where slope is steeper than a certain amount. For example, the highest recommended design slope for gas pipeline projects is 22%, or at most 35% under special circumstances. As a result, crossings a pipeline through areas with steeper slopes will require more excavation. Hence, it is imperative for a gas pipeline designer to identify the slopes of 22-35% and those of higher than 35%.

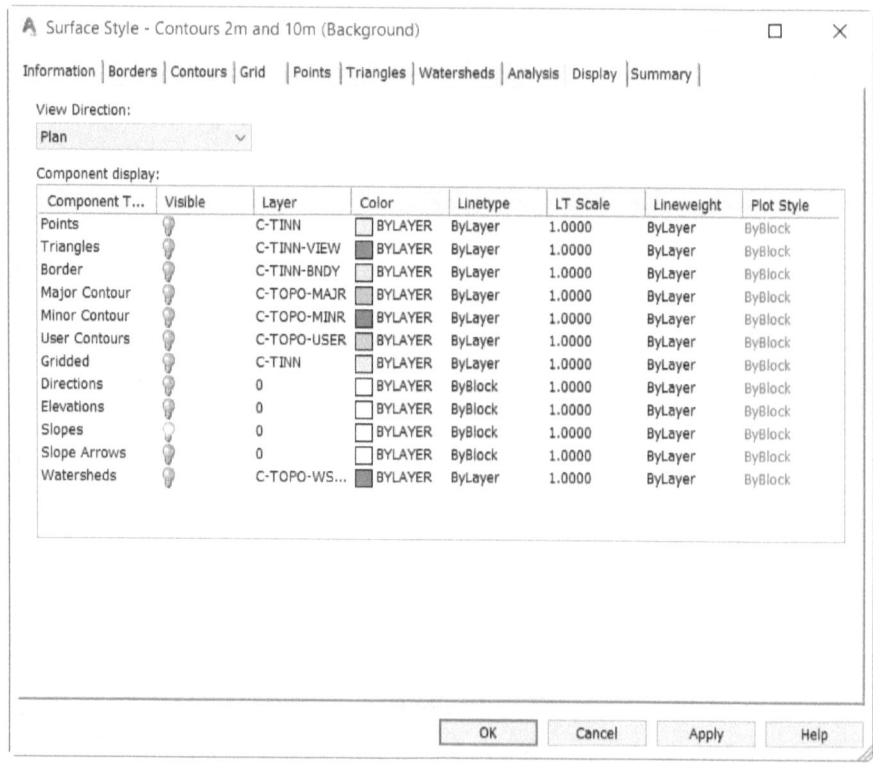

Figure C28

For this exercise, open the file surface-slope.dwg from the folder Project File. In this exercise, the goal is to color the slopes of 22-35% light blue and color the slopes of more than 35% dark blue. For this purpose, open the *TOOLSPACE* window and the *Prospector* tab, right-click on the surface and select *Edit Surface Style*. In the opened window, go to the *Display* tab, turn off all layers except *Slope* (Figure C28) and press *OK*. In the *TOOLSPACE* window, again right-click on the surface and this time select *Surface Properties*. Go to the *Analysis* tab and set the *Analysis Type* option to *Slopes* (Figure C29). In the box called *Ranges*, set the *Number* option to 2 and click on the arrow button to insert the selection into the *Ranges Details* pane.

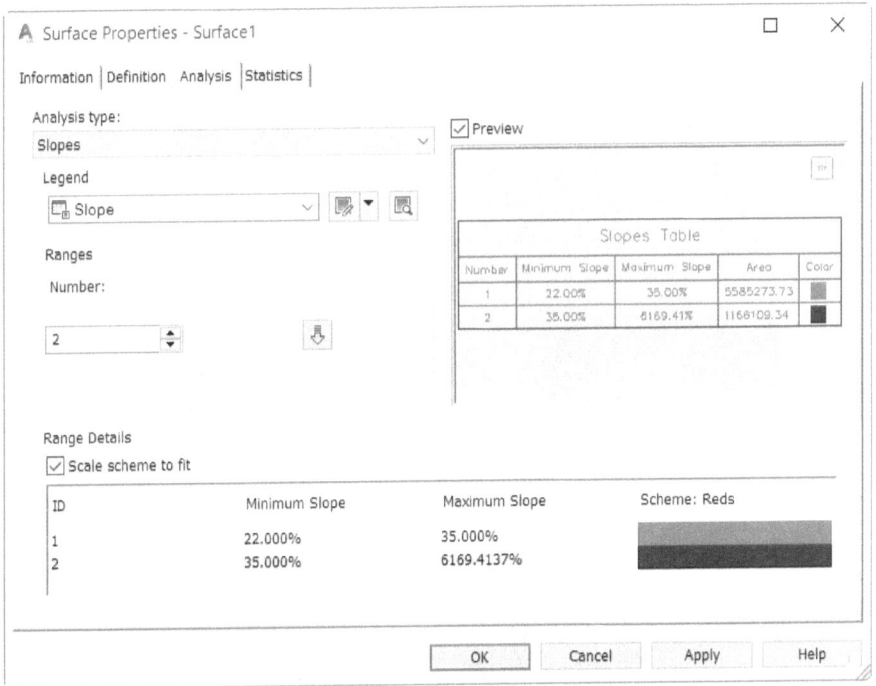

Figure C29

As you can see, this creates two lines in the *Ranges Details* pane. In the first line, type 35 in the *Maximum Slope* column and 22 in the *Minimum Slope* column. In the second line, type 35 in the *Minimum Slope* column but do not change the other column. In the *Scheme* column, select a light blue color scheme for the first line and a dark blue for the second line. After pressing OK, the changes will be applied to the surface. This feature gives designers a better insight into the geography of the project site, and in this example, allows them to choose the pipeline path such that excavation operations are minimized.

36-How to reproduce the points of a surface on a secondary surface based on the elevation of the latter?

In some projects, surveyors are required to prepare a list of point elevations in initial and secondary surveys to facilitate controlling the volume of work done on a point by point basis. In other words, you may have to create a list of points that have the same coordinates but different elevations in initial and secondary surveys, so that they can compared with each other.

To better understand this, you should first learn how to change the elevation of a group of points according to a surface. To learn this, open the file Surface-1.dwg from the folder Project File, which contains a surface without any point. Then, import the file without elevation.txt from the folder Points File.

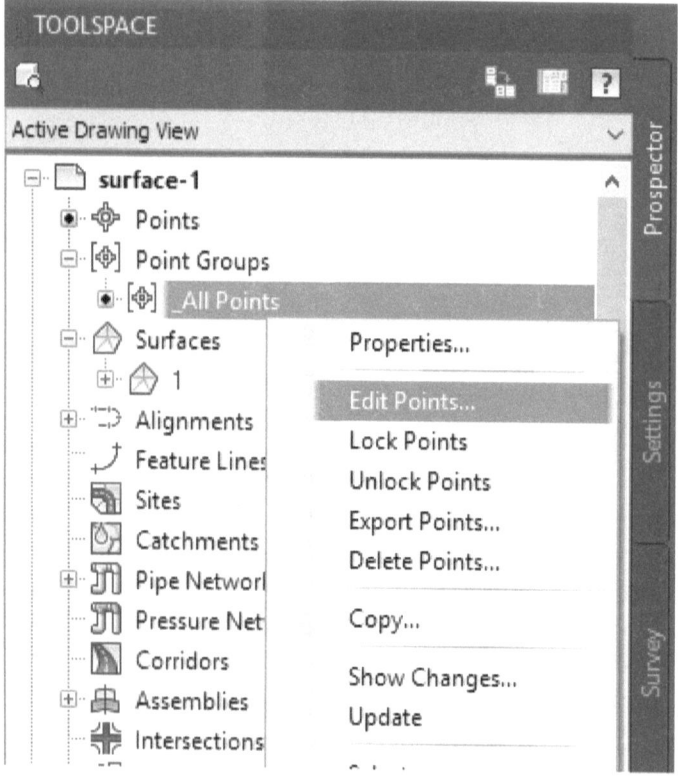

Figure C30

As you can see, all the points of this file have an elevation of zero. Here, the goal is to change the elevation of these points according to the surface. To do this, go the *TOOLSPACE* window and *Properties* tab, right-click on the imported point group and select *Edit Points* (Figure C30). In the opened window, press Ctrl+A to select all the points and then click on *Elevation From Surface* (Figure C31). This opens the window displayed in Figure C32, which asks you to specify the surface from which the elevation must be referenced. After pressing OK, all points will be transferred to the elevation of the surface.

Point Nu...	Easting	Northing	Point Elevati...	Name	Raw Descripti...	Full Descript...	Descripti
5373	0315.2740m	3937.2110m	0.000m		1	1	
5374	0351.6560m	3919.2730m	0.000m		Export...		
5375	0388.2240m	3906.6540m	0.000m				
5376	0423.5750m	3896.9800m	0.000m		Renumber...		
5377	0498.2810m	3895.0880m	0.000m		Datum...		
5378	0318.5660m	3968.9140m	0.000m		Elevations from Surface...		
5379	0351.4420m	3956.2180m	0.000m				
5380	0389.3470m	3944.9150m	0.000m		Delete...		
5381	0426.7820m	3935.4470m	0.000m		Select		
5382	0467.8400m	3930.0900m	0.000m		Zoom to		
5383	0495.6280m	3931.6750m	0.000m		Pan to		
5384	0321.4850m	4005.9710m	0.000m		Lock		
5385	0350.7350m	3991.9420m	0.000m		Unlock		
5386	0384.8800m	3984.4610m	0.000m				
5387	0429.9610m	3972.9420m	0.000m		Export LandXML...		
5388	0455.6220m	3967.3180m	0.000m		✔ Line Shading		
5389	0492.3910m	3962.3680m	0.000m				
5390	0312.2570m	4058.7170m	0.000m		Copy value to clipboard		
5391	0345.7970m	4048.4170m	0.000m		Copy to clipboard		

Figure C31

For example, to reproduce the points of Surface 1 on Surface 2, you must create Surface 1, import the points of Surface 2, change the elevations of the imported points according to Surface 1, extract the transferred points and add them to the points of Surface 1. Then create Surface 2, import the points of Surface 1, extract the points and add them to the points of Surface 2.

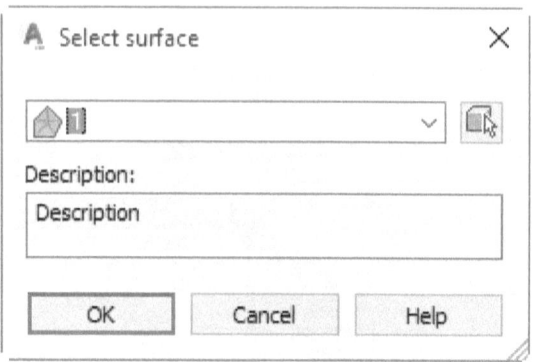

Figure C32

37- How to create a point table based on the elevation difference of two surfaces?

For this exercise, open the file Compare-Point.dwg from the folder Project File. This file contains two point groups and two surfaces, one initial and another secondary, which are named 1 and 2. Here, we first want to create a series of points with elevations equal to the elevation difference of the two surfaces. This can be done by creating a comparison surface. To do this, open the *Surfaces* menu and select *Create Surface*. In the opened window (Figure C33), set the *Type* option to *TIN volume surface* and change the fields to the right of *Base Surface and Comparison Surface to 1 and 2 respectively (names of the surfaces in this exercise). Press OK to close the window and create the comparison surface. The elevation of each point on this comparison surface is equal to the elevation difference of the two input surfaces at the same coordinates. Note that the output of this process is the comparison surface itself rather than the countless points that the software has to calculate internally to obtain this* surface.

To generate these points, go to the *Toolbox* tab in the *TOOLSPACE* window, expand the *Report Manager* branch and the *Surface* sub-branch, right-click on *Surface Points to CSV* and select *Execute* (Figure C34). This opens the window shown in Figure C35.

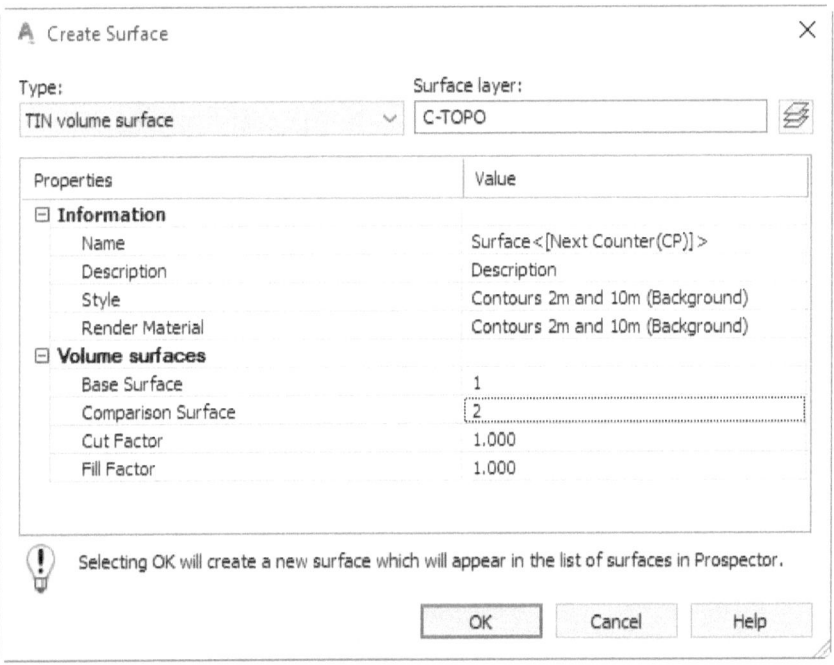

Figure C33

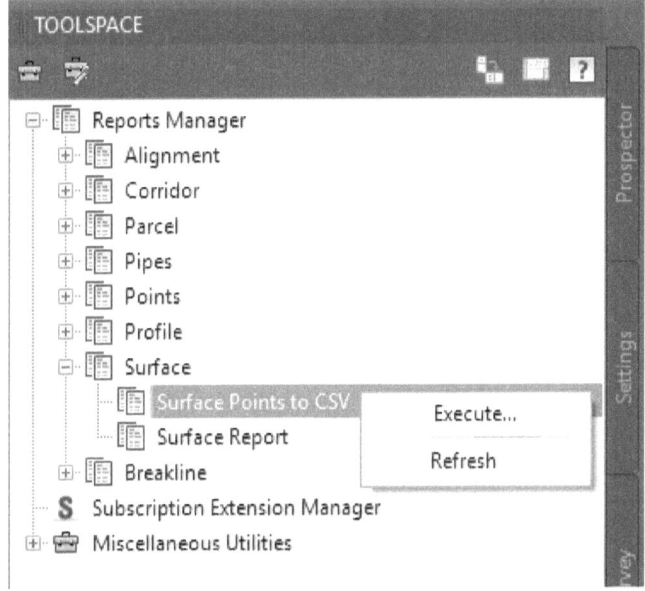

Figure C34

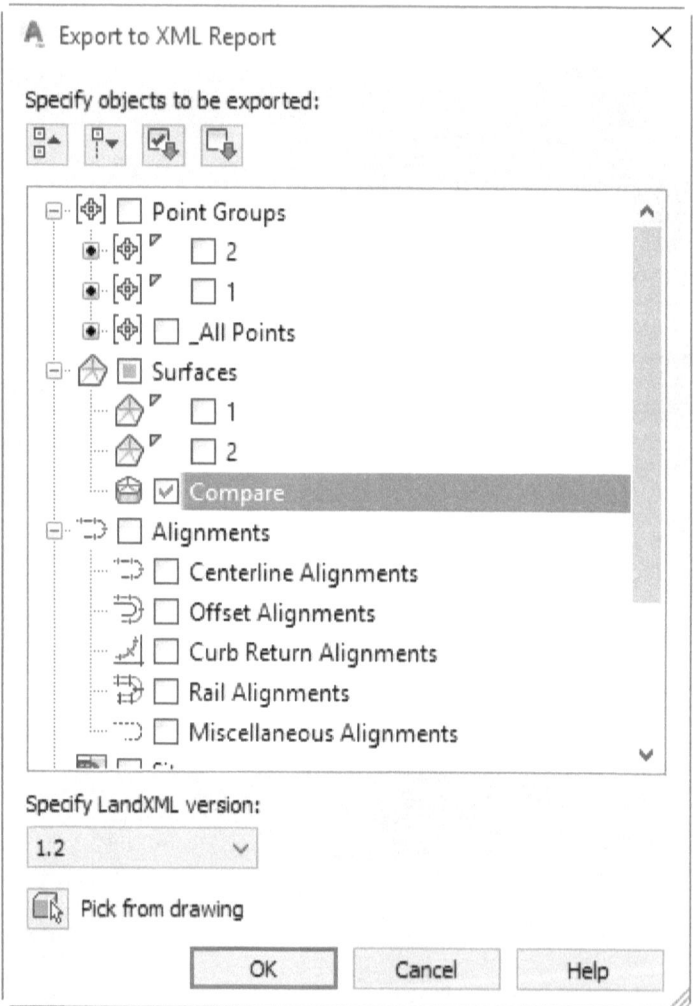

Figure C35

In this window, uncheck all the boxes except the one related to the comparison surface, and press OK to export the surface as a CSV file (where the coordinates of each point are displayed alongside the corresponding elevation difference value).

You can use Excel or similar software to convert this file into a txt file.

38- How to display the bend angle and the number of bends over the PIs of a pipeline?

In pipeline construction projects, the number of pipe pieces that must bent has important cost and implementation implications and need to be carefully considered during preliminary studies and design.

In the case of gas pipelines, there are regulations that restrict the maximum bend angle that can be applied to a pipe of certain diameter.

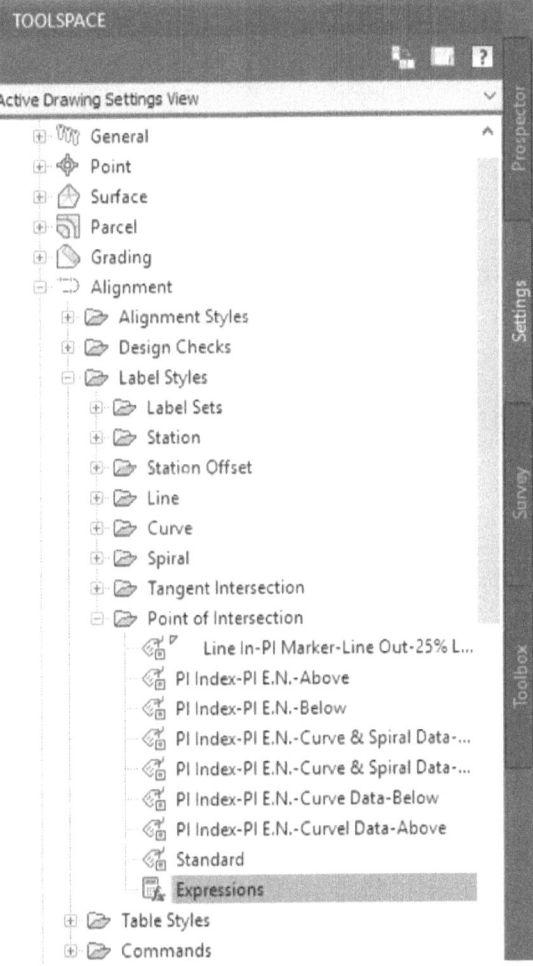

Figure D1

For example, for a 30-inches pipe, the angel of bend must not exceed 16°.

As a result, when the angle of PI (point of intersection) is less than or equal to 16°, the bend can be constructed with a single 12 meter pipe, but when this angle is 17°, the bend must be applied using two pipes. To maintain control over the implications of this issue, it is preferable to make the bend angle appear alongside the PIs.

Even better, you can define a relationship for dividing this angle by 16 degrees (depending on the regulation) and rounding up the answer and make the label show the number of pipe pieces needed at each PI.

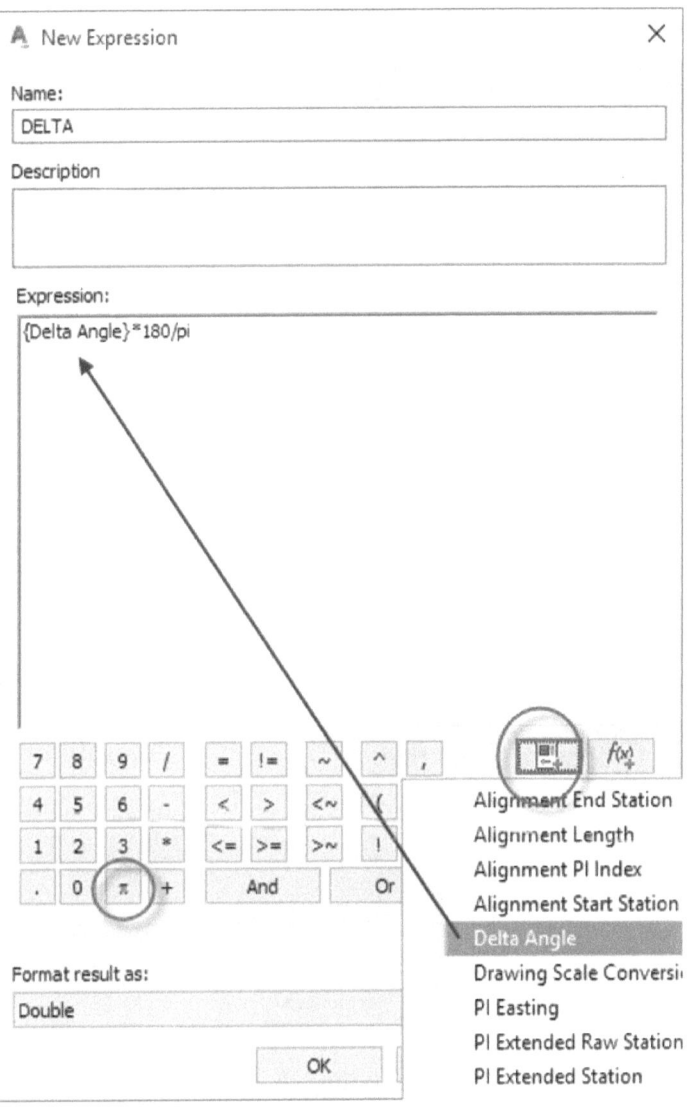

Figure D2

For this exercise, open the file DETA-BEND-LABLE.DWG from the folder <u>Project File</u>. This file is a drawing of a 30-inch gas pipeline. Here, the goals is to make the bend angle and the number of bends appear over the PIs. After opening the file, go to the *TOOLSPACE* window and the *Settings* tab and expand the *Alignment* branch, and *Label Styles* and *Point of Intersection* sub-branches (Figure D1). Right-click on *Expressions* and select *new*. In the opened window, first type a name for the expression to be created, then use the function button below to add the *Delta Angle* function (Figure D2). Multiply the angle by 180 and divide it by π to convert it to degrees.

Figure D3

Save the Expression by pressing the OK button.

Next, you need to create another expression, as described above, but this time for the number of bends. In this expression, the previous expression, i.e. the bend angle, must be divided by 16 and then rounded up. For example, if the bend angle is 18 degrees, first it will be divided by 16 to become 1.125, and then will be rounded up to become 2, which is the number of bends needed at that particular PI. To define this new expression, first name the expression, then click on the function button below and select the *ROUNDUP* function (Figure D3). Inside this function, the bend angle expression you just made must be divided by 16. To do this, click again on the function button and select the bend angle expression you made (we named it DELTA), and then type /16 to divide it by this number. Before closing the *ROUNDUP* function, you must type a comma followed by the number of decimal places to which the value inside the parentheses will be rounded (in this exercise, it is 0). Save the expression by pressing OK.

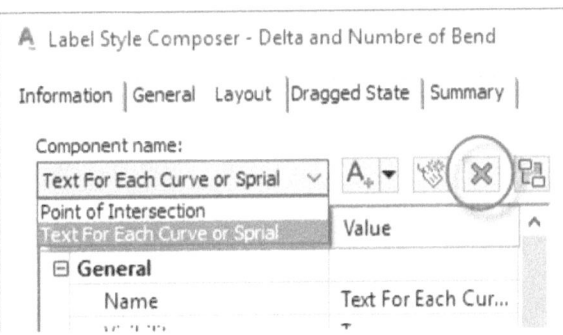

Figure D4

Now you have to define a *Label Style* consisting of these two expressions. For this purpose, follow the path explained earlier for reaching Expressions (Figure D1), but this time click on Point of Intersections and select New. In the opened window, first type a name in the Information tab (here we named it Delta and Number of Bends), then press the button marked in Figure D4 to remove the two previous components and use the button shown in Figure D5 to create a new component and name it.

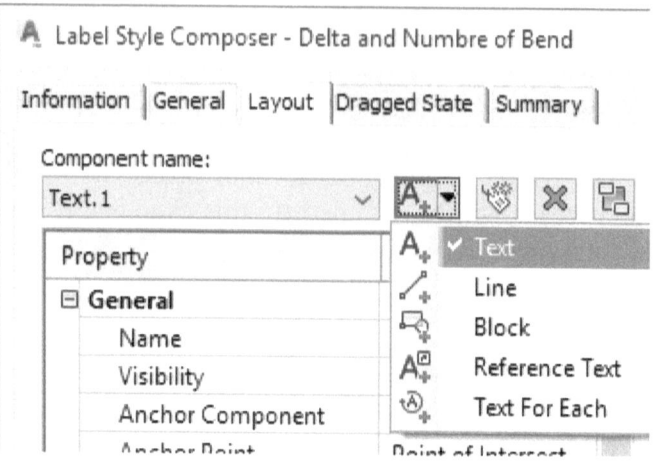

Figure D5

After creating the component, it must be assigned with the created expressions.

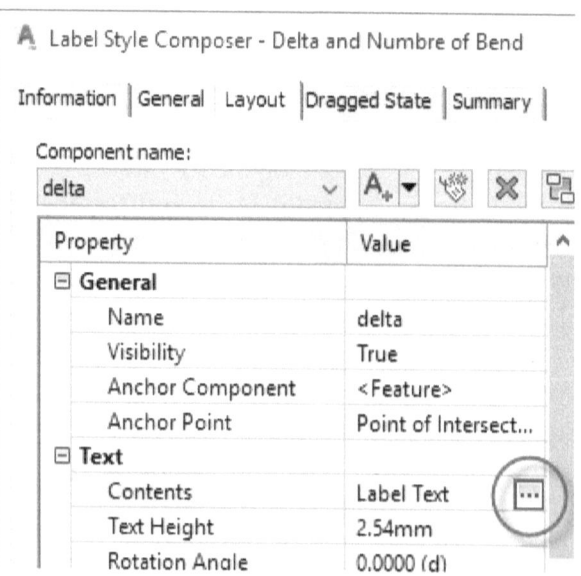

Figure D6

For this purpose, click on the box in front of *Text* (Figure D6) to open the *Text Component Editor* window shown in Figure D7.

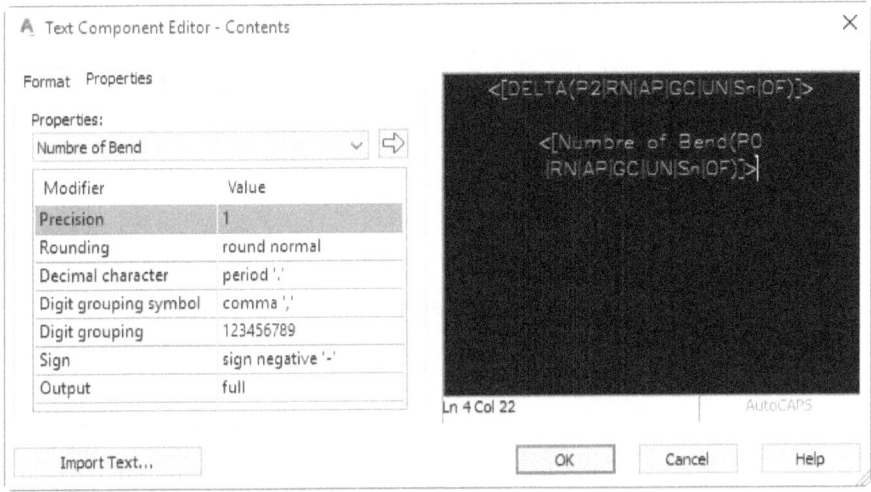

Figure D7

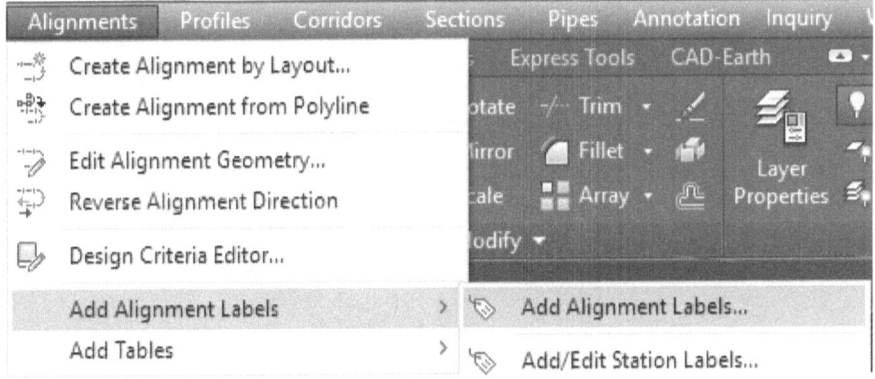

Figure D8

In this window, delete the contents of the right pane, open the Properties drop-down menu and select one of the created expressions, specify the number of decimal places in the *Precision* field (2 for the bend angle and 0 for the number of bends), and press the arrow button to add the selection to the right pane. After performing this process for both of the created expressions, press OK to close the window. You just created the label with the desired definition, and just have to make the labels appear over the PIs.

Figure D9

To insert the labels into the drawing, open the *Alignments* menu, click on *Add Alignment Labels* and select *Add Alignment Labels...* (Figure D8). This opens the window shown in in Figure D9. In this window, set the *Feature* menu to *Alignment* and the *Label type* menu to *Multiple Point of Intersection*. Find and select the label style created in the previous step in the *Point of Intersection label style* menu and press the *Add* button and click on the desired alignment to make the labels of the bend angle and the number of bends appear over the PIs (Figure D10).

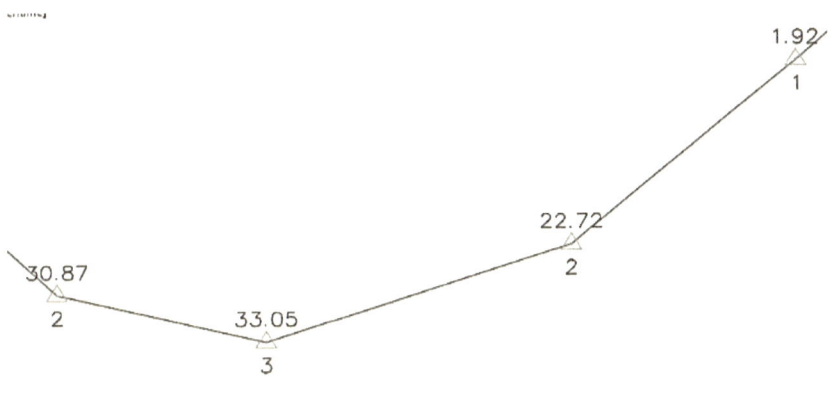

Figure D10

As you can see, in this example, the fourth PI has a bend angle of less than 16°, and therefore requires a single bend, the first and third PIs have a bend angle of between 16° and 32° and require two bends, and the second PI, which has a bend angle of greater than 32°, needs three bends.

39- How to calculate the coordinates of a point based on the station and offset?

For this exercise, open the file DETA-BEND-LABLE.DWG from the folder Project File, which is a drawing of a gas pipeline construction project. Assume that during the operation, the project officers inform you of the presence of a power pole at the station 18+430 with 53 meters offset to the left of the pipeline axis, and ask you to use this data, namely the (station) chainage and the offset of the power pole, to add this feature to the existing drawings and extract its coordinates for obtaining the necessary permits. To do this, open the *Points* menu, select *Alignment*, and then select *Station/Offset* (Figure D11). The software will ask you to choose the alignment by clicking. After choosing the alignment, the message "*Specify station along baseline*" will appear on the command line. In response to this massage, you must type the target station, i.e. 18+430, and press enter.

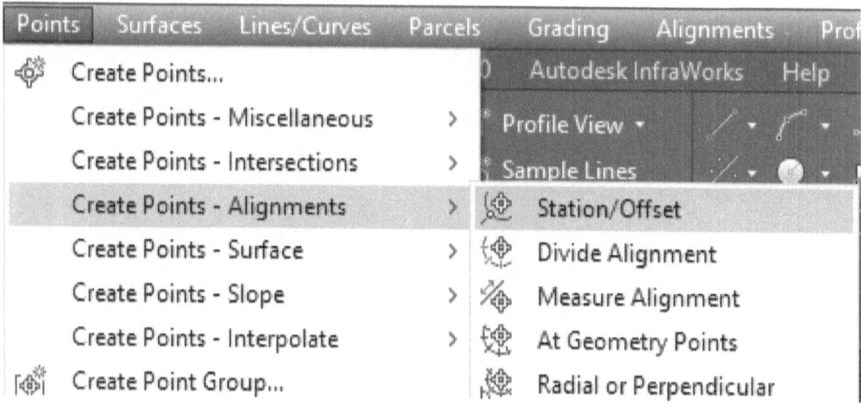

Figure D11

In response to the next message (Specify an offset <0.000:), you must type the target offset, i.e. -53 and press enter (negative sign is used because the target is on the left side of the alignment).

You must then specify in the command line a code and an elevation for the point to be created. After these steps, the software will create a point on the specified station and offset. You can extract the coordinates of this point with the ID command. For this exercise, the coordinates of the point should be:

X = 590819.294 Y = 3943742.659

40- How to determine the station (chainage) and offset of a point based on its coordinates?

For this exercise, open the file DETA-BEND-LABLE.DWG in the folder Project File. This file is related to a gas pipeline construction project. In the case of gas pipelines, the length of buffer zone may vary depending on the pipe size and thickness. For example, the pipeline of this project has a 35 meter wide buffer (from the pipeline axis to each side) between the stations 15+000 and 16+500, which increases to 42 meters from the station 16+500 until 21+300. Assume that you have used a handheld GPS to obtain the coordinates of a power pole in the vicinity of pipeline (X = 589450 Y = 3945199) and now want to determine whether this pole is within the pipeline buffer zone.

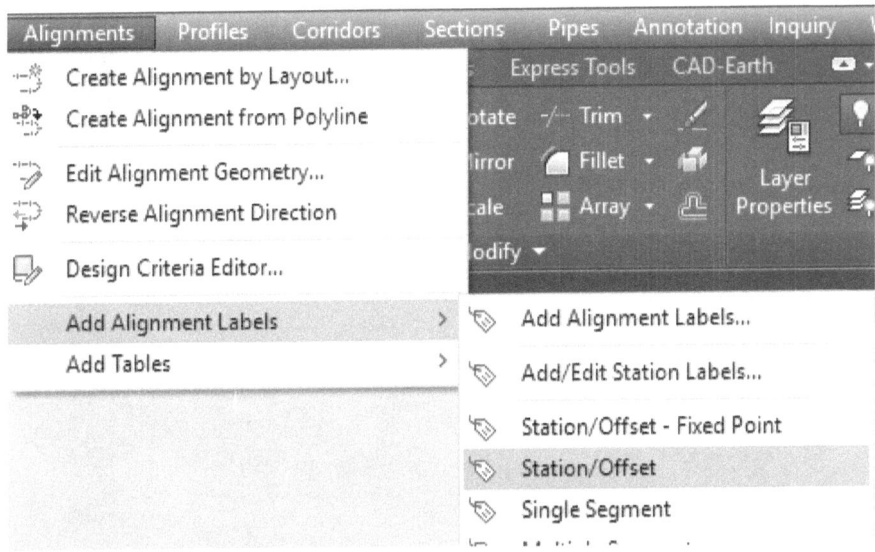

Figure D12

For this purpose, first create a point at the given coordinates (follow the instructions for creating a point or a circle at specified coordinates). Then open the *Alignments* menu, select *Add Alignment Labels* and then *Station/Offset* and click on the point to insert the label (Figure D12). As you can see, this point is located at the station 16+300.67 at an offset of 32.37 meters, which means the pole is within the pipeline buffer zone.

41- How to calculate the station (chainage) and offset for a point coordinate file?

Assume that in a road construction project, a road with a buffer width of 35 meters on each side has to pass through a private orchard. To conduct a financial analysis, all trees of this orchard are indexed and the exact coordinates of every tree are obtained by surveying. Now, the goal is to make a list of tree indexes that shows which tree is within the road buffer zone.

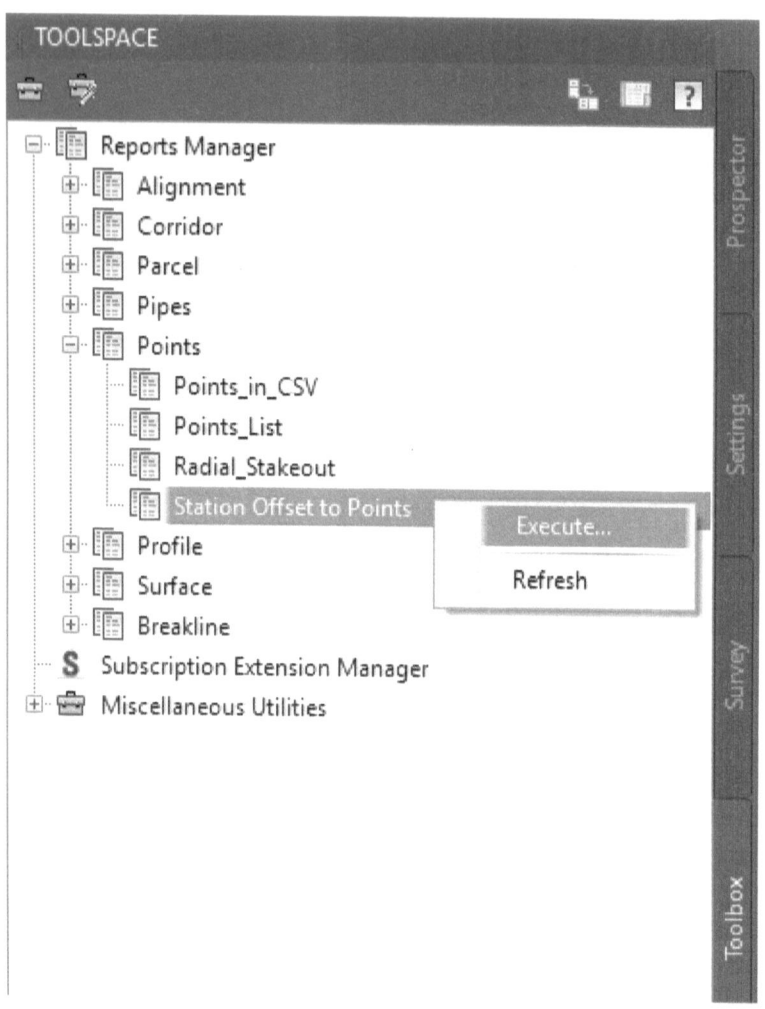

Figure D13

For this exercise, open the file <u>alignment-buffer.dwg</u> from the folder <u>Project File</u>. As you can see, this file shows a road passing through an orchard with 314 trees. To check the position of the trees with respect to the road buffer, you have to create a file for comparing the offset of each individual tree with the buffer. To do this, open to the *TOOLSPACE* window and the *Toolbox* tab, expand the *Report Manager* branch and the *Points* sub-branch, right-click on *Station Offset to Points*, and select *Execute*. In the opened window, select the points and the alignment and then type a name and a path for the output file to be created. This output file will contain a chainage (station) and an offset for each point, which can be easily compared with the considered buffer length. In this example, any tree with an offset of less than 35 or -35 will be within the road buffer zone, and should be considered in the calculation of compensations.

42- How to draw trench lines for area calculation and cadastral mapping based on cross sections?

One of the common requirements of road construction and cadastral mapping projects is to calculate the area of excavation in each segment, which is a function of trench lines and cross sections. When design and site corridor are available, the corridor lines can be easily extracted. But this cannot be done as easily when only the drawing files of the route axis and the longitudinal and cross sections are available. This is because then you need to calculate the width of excavations at both left and right side of the axis at those stations for which cross sections are available, draw the points at resulting offsets, and finally connect the points to obtain the corridor lines.

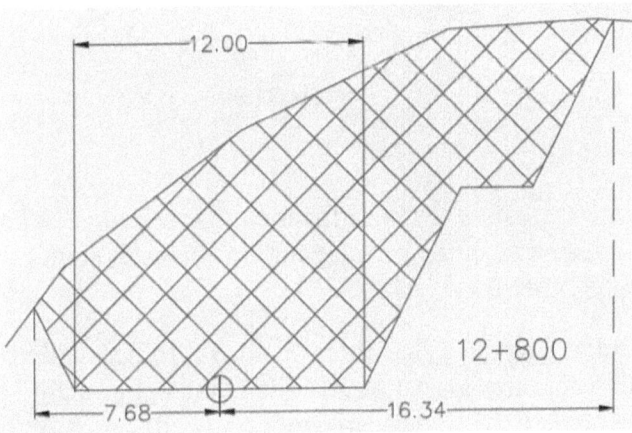

Figure D14

For example, consider Figure D14, which shows a cross section drawn for a road with a width of 12 meters. It can be seen that although the road is only 12 meters wide, the excavation width is 7.68 meters on the left and

16.34 meters on the right. Obviously, drawing this diagram at regular cross sections and creating points according to the offsets obtained at each station will be very time consuming. In the software however, this process can be done more conveniently by using the *Create Point Alignment* and *Import From File* features to create a point file in a specific format and import them into the drawing.

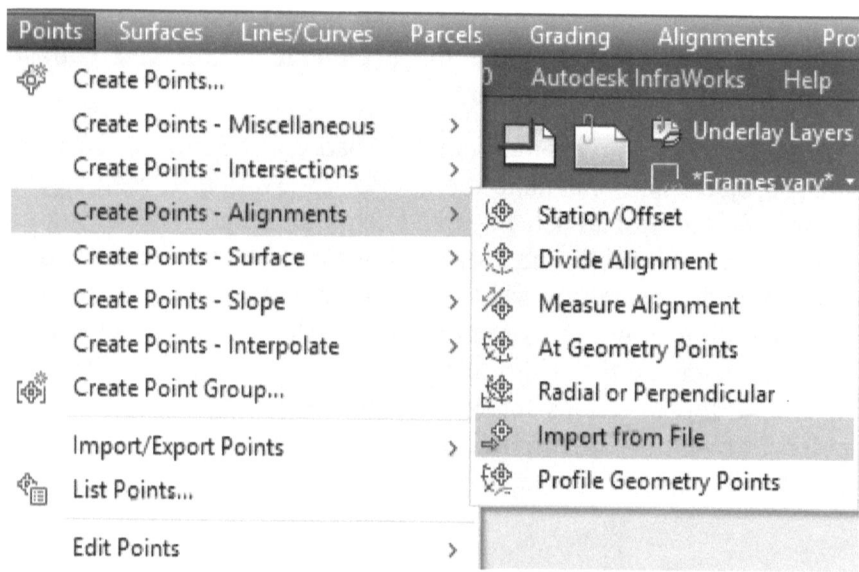

Figure D15

First you must create a text file for the cross sections. When doing so, you need to create two lines of information for each station: one for the left offset and another for the right.

In the folder Project File, there is a subfolder called alignment-cadastre, which contains an AutoCAD file named alignment-cadastre.dwg and a text file named alignment-cadastre-points.txt.

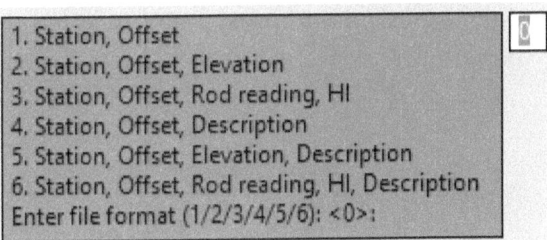

1. Station, Offset
2. Station, Offset, Elevation
3. Station, Offset, Rod reading, HI
4. Station, Offset, Description
5. Station, Offset, Elevation, Description
6. Station, Offset, Rod reading, HI, Description
Enter file format (1/2/3/4/5/6): <0>:

Figure D16

Open the dwg file. As you can see, this file contains a route (alignment) and related cross sections from the station 00+000 to 00+156.40. The goal of this exercise is to draw the cadastral maps of this route in the said segment. The text file contains the data of trench lines, which have been extracted from the cross sections and exported as a text file. To determine the area of excavation in the segment of interest, open the *Points* menu, click on *Create Points - Alignment* and select *Import From File* (Figure D15). In the opened window, go to the above-mentioned directory and select the text file. The software then show you a list of standard data formats and asks you to specify the format of data contained in the text file (Figure D16). Since the text file of this exercise is in *Station, Offset* format, here you have to type 1 and press enter. The software then asks you to specify the separator of the columns in the file, which in this particular file is comma, so you need to type 2 and press enter.

The software then asks that what should be done if the station and offset values are incorrect. Typically, it is recommended to replace incorrect values with a greater number, so that errors stand out. Finally, click on the alignment of interest to import the points along that alignment.

43- How to reverse an alignment?

Suppose that after spending some time drawing a project, you notice that the direction of a particular alignment is defined incorrectly and should be reversed. This can be easily done by opening the *Alignments* menu selecting *Reverse Alignment*. Alternatively you can select the alignment, go the *Modify* panel in the ribbon, select *Reverse Direction*, and then click on the alignment and press Enter to see it reversed (Figure D17).

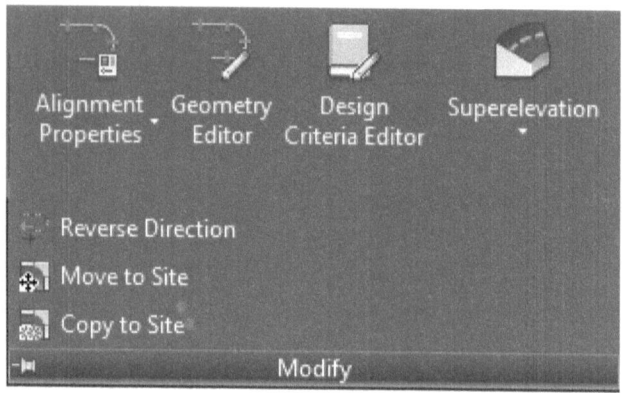

Figure D17

44- How to define alternative stations for an alignment segment?

In road and pipeline construction projects, it is not uncommon to see designers consider two or more alternative routes to circumvent geographical, technical or financial problems. Also known as variants, these alternative routes split from the main route at one point and rejoin it at another point further along the path. Naturally, such variants could be longer or shorter than the main route at that segment.

Since it is typical to prepare the project lines, longitudinal and cross sections, volume tables, soil classification tables, etc. for the entire length of an alignment, it is unreasonable to change or transform all the parameters only because a variant has changed the length of a segment of the project. A more convenient solution is to define the affected alignment segment separately with alternative stations.

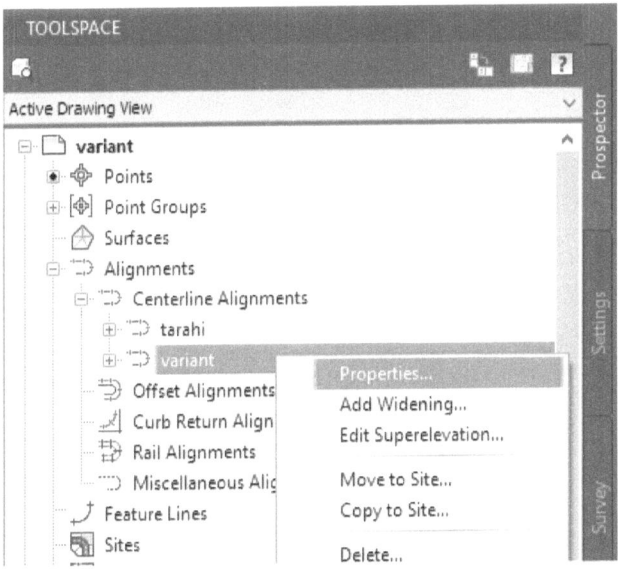

Figure D18

For this exercise, open the file <u>Variant.dwg</u> from the folder <u>Project File</u>. The route drawn in this file has two branches, the lower branch is the main route and the upper one is a variant, which connects the station 0+500 and 4+300, but with 858.15 meters increased length. Going through this alternative route will change the chainage of the route at the second intersection from 4+300 to 5+158.15. To ensure correct calculation of stations from the end of the variant, the station 5+158.15 of the variant should be defined equal to the station 4+300 of the main route. To do this, open to the *TOOLSPACE* window and the *Prospector* tab, right-click on the variant alignment (in this exercise this alignment is named *Variant*), and select *Properties* (Figure D18). In the opened window, go to the *Station Control* tab and press the button shown in Figure D19 and click on the 5+158.15 station to add an station equation. To complete the equation process, change the value under the *Station Ahead* column to 4+300 and set the *Increase/Decrease* column to *Increasing* (see Figure D19). Press OK to close the window. After this change, the station after the second intersection become equal.

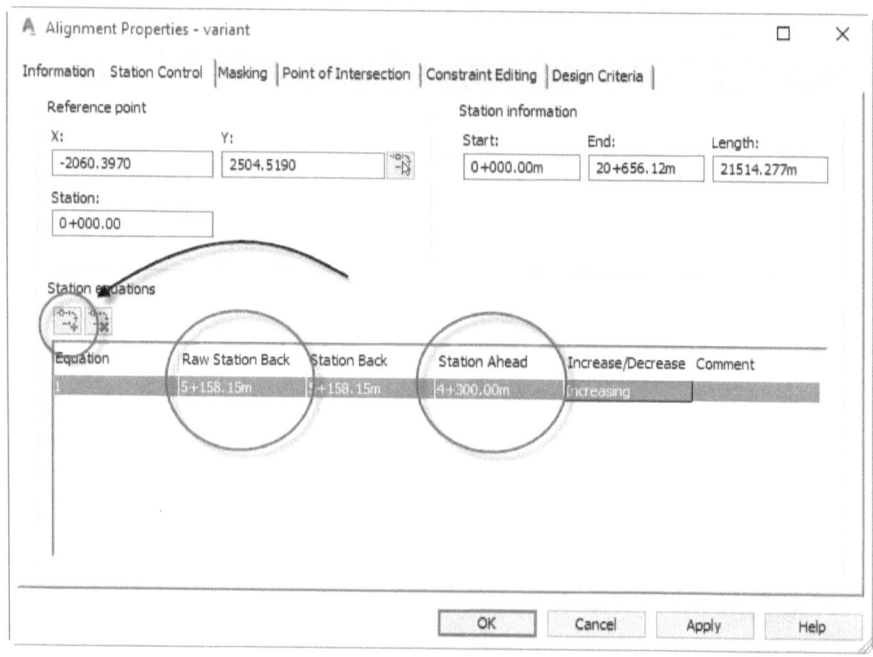

Figure D19

45- How to create station label exclusively for one alignment segment?

To create a station label for just one segment of a alignment, open the *Alignments* menu, select *Add Alignment Labels* and then *Add / Edit Station Labels*, and click on the alignment (Figure D20) to open the window illustrated in Figure D21. In this window, uncheck the boxes, type the start and end station of the segment of interest in the *Start Station* and *End Station* columns respectively, and press OK to apply the changes.

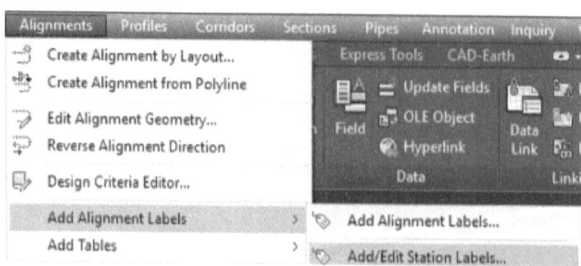

Figure D20

Figure D21

46- How to customizes the station labels without altering the reference point?

Suppose that the reference point of an alignment has been moved by 1500 meters, and now you want to add this 1,500 meters to the station labels without actually altering the reference point. Alternatively, suppose that in exercise 20, you want to make the labels subtract the 858.15 meters that is added the route length after the station 4+300 without changing the actual station values.

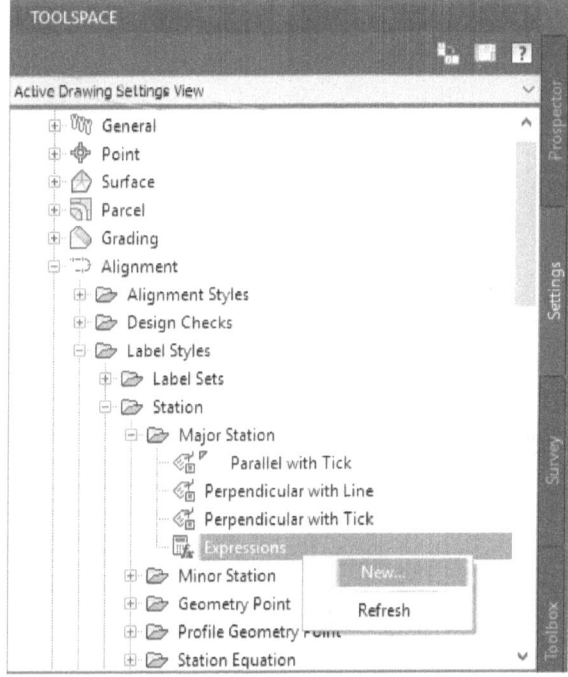

Figure D22

Both of these label customizations can be simply done by creating appropriate expressions and applying them on the labels of interest. For this purpose, go to the *Settings* tab in the *Toolspace* window and expand the *Alignment* branch, and then *Label Style* and *Station* sub-branches (Figure D22). Right-click on *Expressions* under the *Major Station* sub-branches and select *New* to open the window shown in Figure D23. In this window, type a name in the *Name* field, then press the function button shown in Figure D32 and select the *Station* parameter, Finally, type +1500 in front of the inserted function and press OK to close the window.

Figure D23

After creating the expression, you need to introduce it to a label style, and then apply that label style to the desired alignment. To introduce the expression to the label style, you can edit one of the existing label style or create a new label style altogether. To create a new label style, follow the path explained earlier to find *Major Station* (Figure D22), right-click on *Major Station*, and select *New* to open the window shown in Figure D24. Type a name in the *Information* tab of the opened window, then go to the *Layout* tab and click on the field to the right of *Contents* (see Figure D24) to open another window (Figure D25).

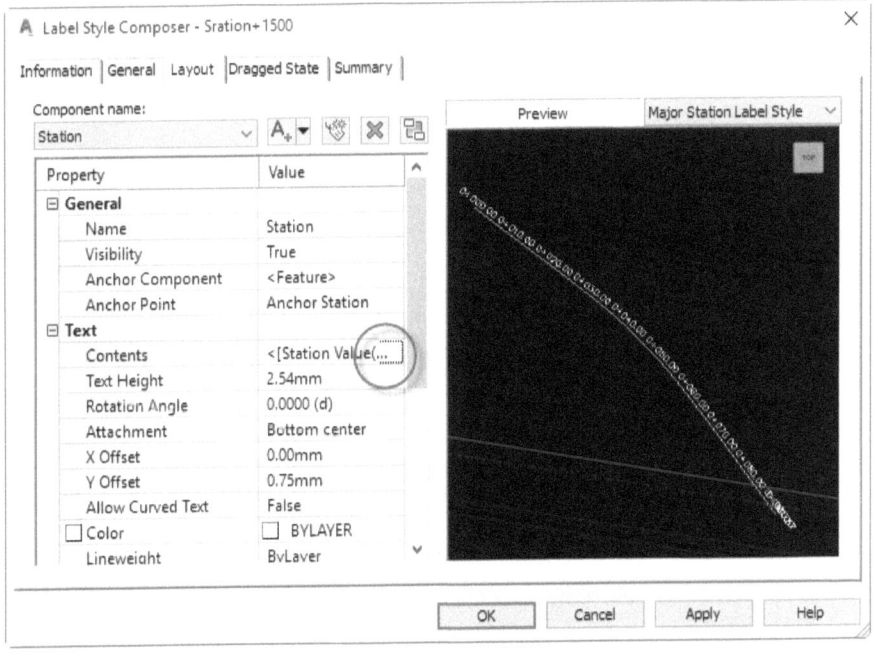

Figure D24

In this window, delete the content of the right pane and then browse the *Properties* menu and select the expression you just created (we named it Station+1500). Click on the arrow button to move the selected expression to the right and press *OK* to close the window.

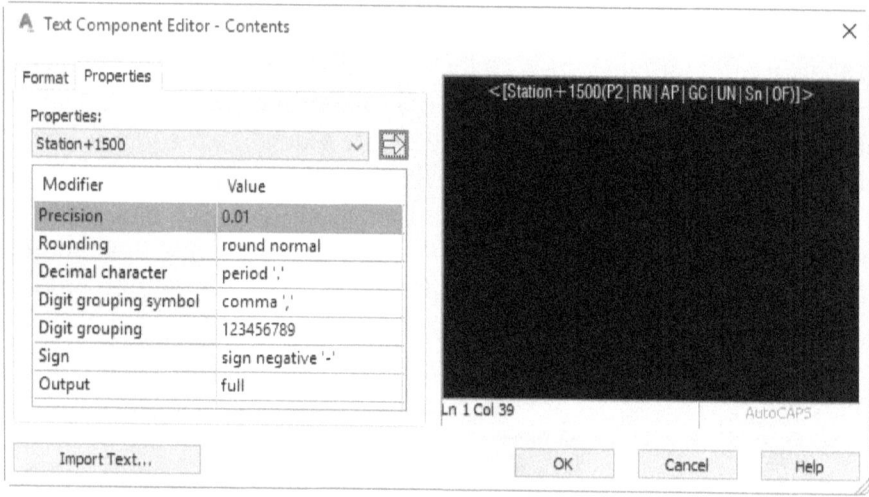

Figure D25

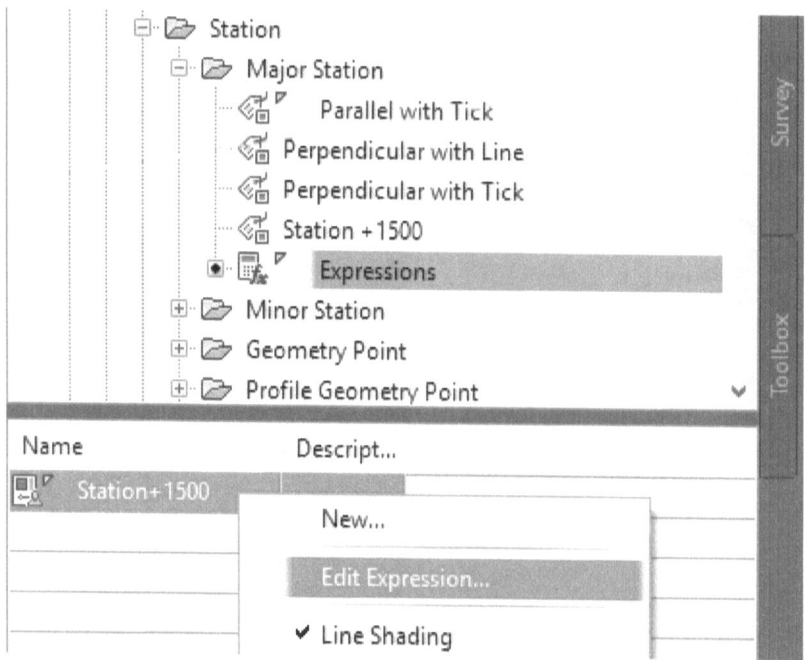

Figure D26

Note: In order to make adjustments such as applying the + sign for splitting the station values, you must create the expression in the *Station* format. If you have not done so, you must click on *Expression* at the same window and right-click on the expression of interest in the list below and select *Edit Expression* (Figure D26). In the bottom of the opened window, change the format of the expression to *Station* (Figure D27).

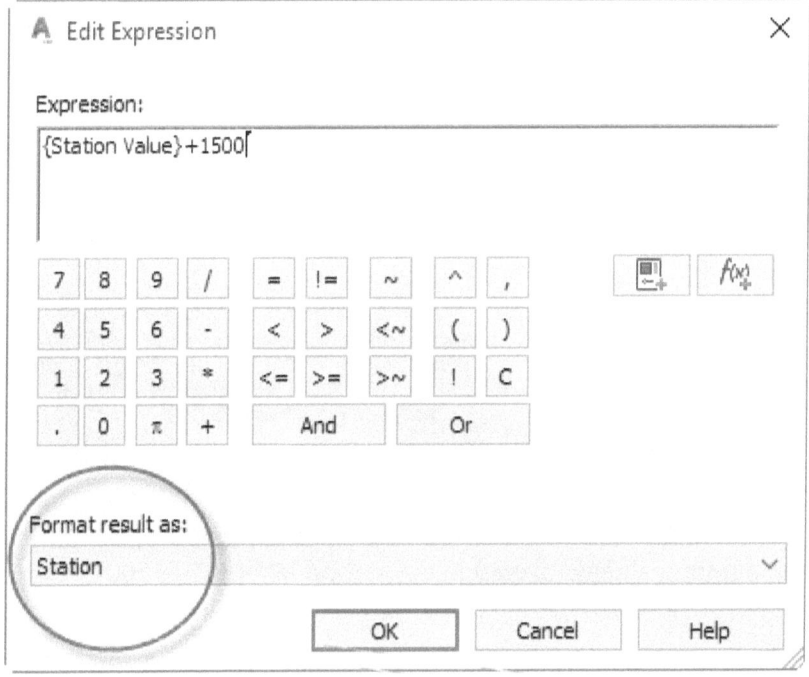

Figure D27

After creating the label style and assigning it with an expression, you must make the label appear on the alignment. To do this, open the *Alignments* menu, select *Add Alignment Labels* and then *Add/Edit Station Labels*, and click on the alignment (Figure D20). This opens the window shown in Figure D28.

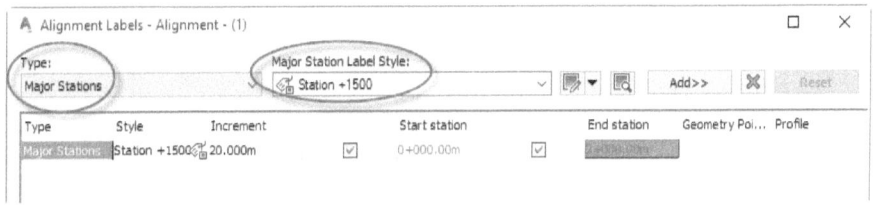

Figure D28

In this window, set the *Type* menu to *Major Station* and find the label style you just created in the *Major Station label style* menu, (in this exercise, we named it Station + 1500), and press the *Add* button. Once finished, press the OK button to apply the changes.

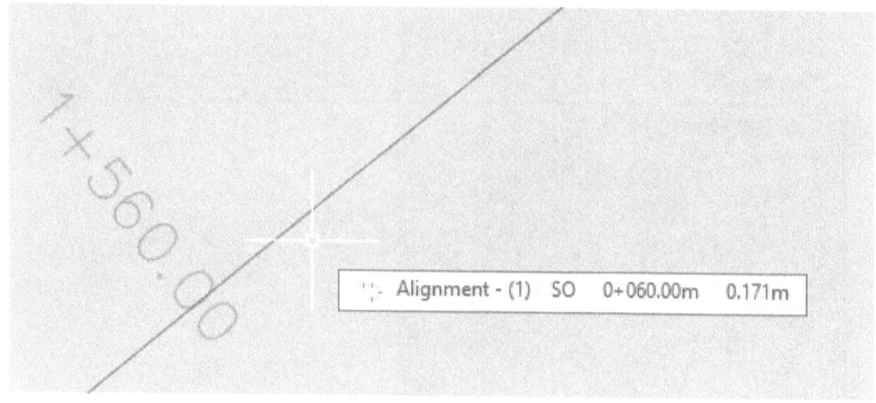

Figure D29

In Figure D29, you can see that the mouse marker is on the station 00+060, but the value displayed in the station label is 01+500.

47- How to adjust the number of decimal places and the position of + sign in the station label?

Many Civil3D users often struggle to understand that why station values are displayed without decimals or with too many decimal places and why these values are split by their hundreds instead of thousands. For example, the station 1525 could be displayed as 15+25 instead of 01+525, or with unnecessary decimal places such as 1+525.000.

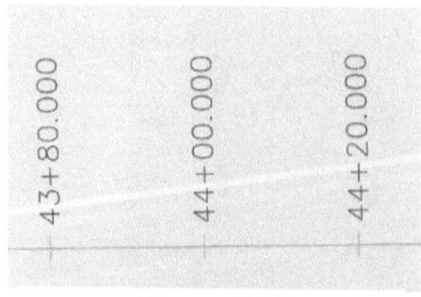

Figure D30

For this exercise, open the file Edit-Labels.dwg from the folder Project File. This file is a drawing of a route with station labels shown in Figure D30. In this example, a station value such as 4+400 is displayed as 44+00.000, which is not only split by its hundreds place (instead of its thousands place) but also has three unnecessary decimal places.

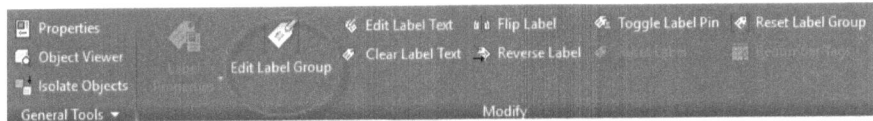

Figure D31

To fix this problem, select the existing labels and click on the *Edit Label Group* button in the ribbon (Figure D31) to open the window shown in Figure D32.

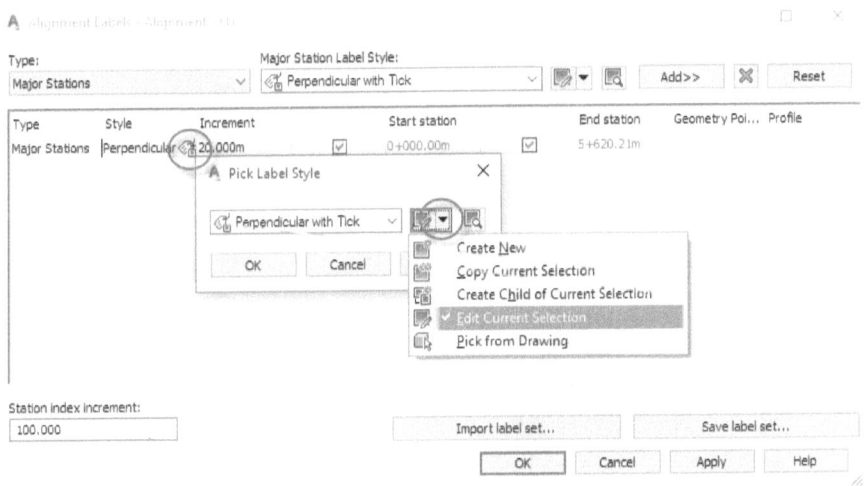

Figure D32

In this window, click on the icon under the *Style* column (see Figure D32). In the opened window, click on the cascading menu shown in the figure and select *Edit Current Selection*. This opens the window shown in Figure D33.

In this window, click the field in front of *Contents* (marked in the figure) to open the window shown in Figure D34. In this window, delete the contents of the right pane, set the *Properties* menu to *Station Value*, type 1 in the Precision field (this means no decimal places will be displayed), and set the *Station character position* to 1+000. Press the arrow button to move the selected property to the right pane, and finally press OK to close all opened windows.

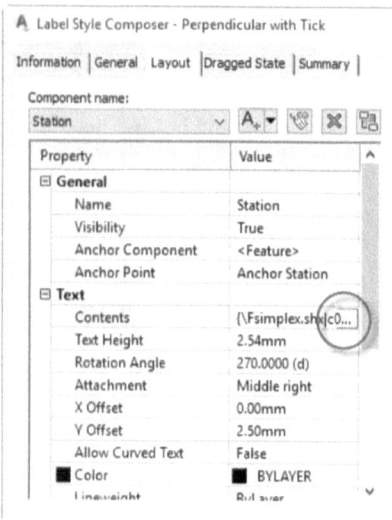

Figure D33

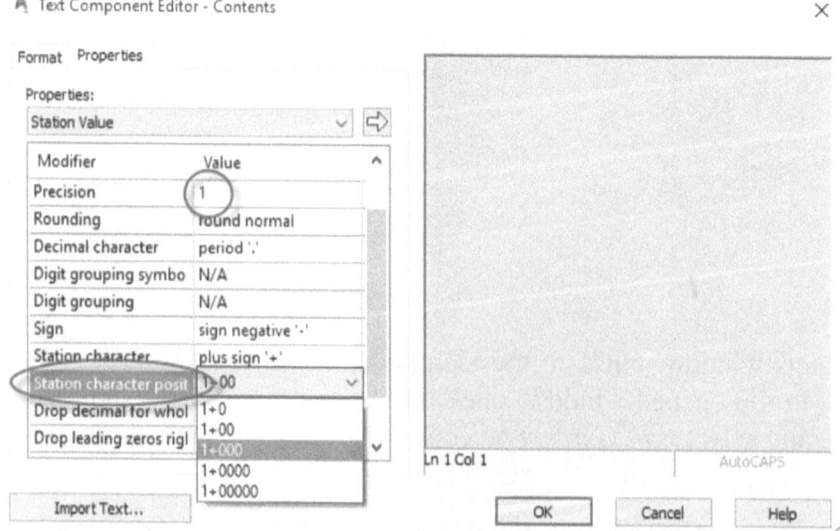

Figure D34

48- How to create station labels for the stations included in the volume table without a corridor?

In the design stage, when volume table, corridor, and alignment are linked, you can create station labels for any station that is included in the volume table using the CIVIL3D features that will be described in later exercises. But during the construction of a given plan, when you only have the alignment, volume table, and list of sections, without them being linked to the corridor, the software does not have a specific option for this purpose. However, you can use a simple trick to achieve the same result.

CIVIL3D is able to draw a longitudinal profile from a text file. The text file to be used for this purpose should consists of station and elevation values separated by a space. For example:

0 1001.12

12.18 1003.18

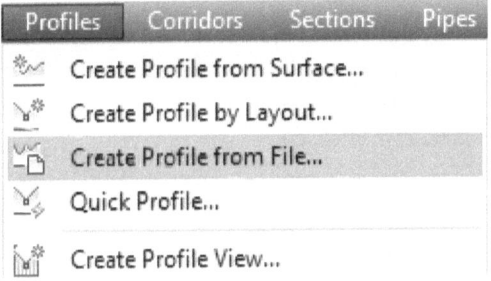

Figure D35

For this exercise, we use an excel file, a drawing file, and a text file contained in the subfolder <u>label from file</u> in the folder <u>Project File</u>. The drawing file is a simple project consisting of a single alignment. The excel file is a list of volumes related to that alignment. If you open this file, you will find that the volumes given in this file belong to non-round sections positioned at irregular intervals. The goal is to create labels on the alignment at those stations for which volume table contains information. As mentioned, you can do this by creating a longitudinal profile with the proper format. For this purpose, create another Excel file and insert the station values in the first column.

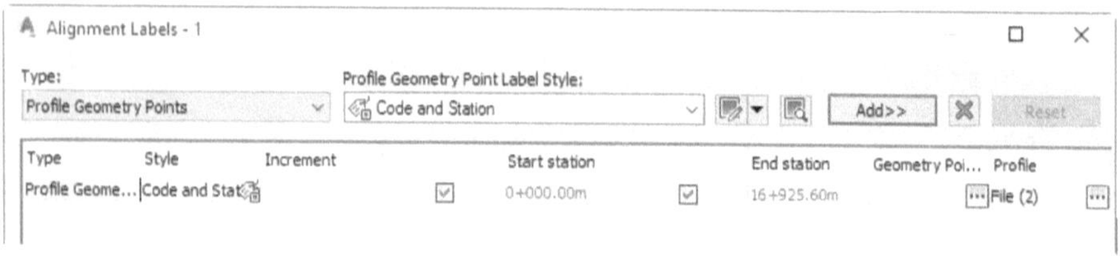

Figure D36

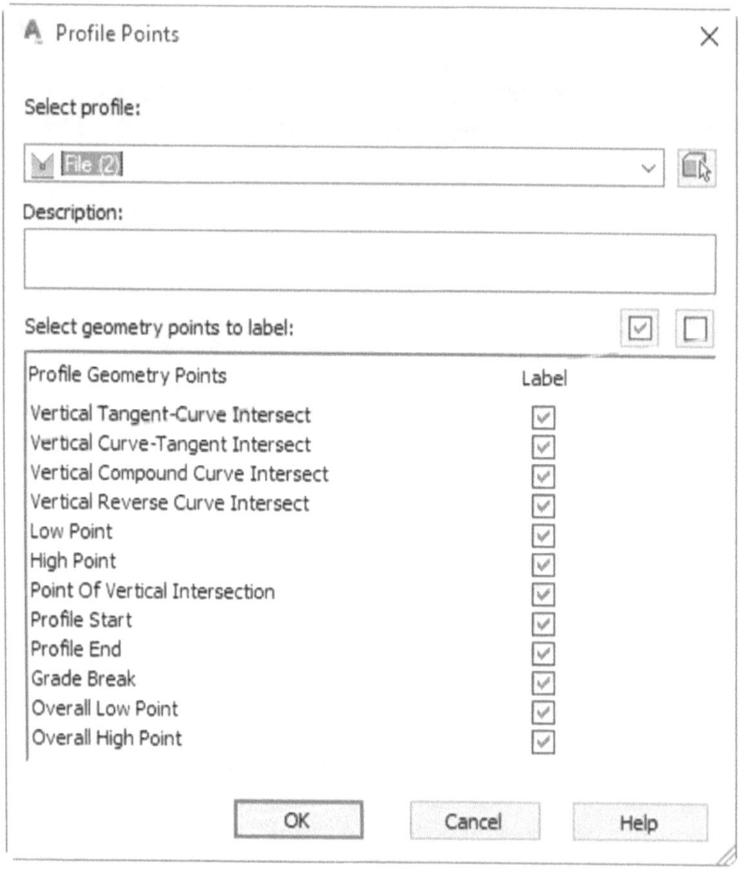

Figure D37

Normally, the second column should contain the elevation values of the profile, but since here the elevation is unimportant, you can fill the entire second column with zeros. Alternatively, you can insert the volume values in the second column to make them appear alongside the corresponding stations. Save the Excel file in the text (tab delimited) format. Open the text file and use Ctrl+H to convert all Tabs to Space and save the changes as a text file such as Label.txt.

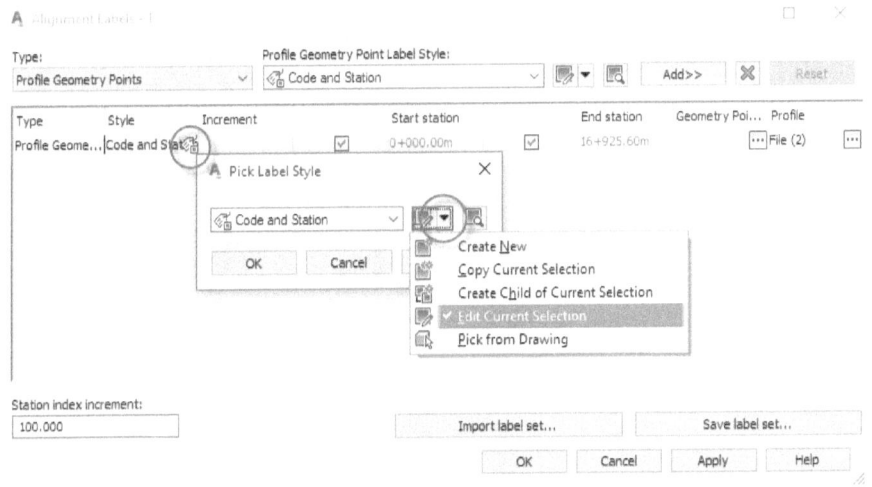

Figure D38

Next, open the drawing file and start creating a profile from the text file. To do this, open the *Profiles* menu and select *Create Profile from File* (Figure D35). In the opened window, select the text file you created. In the next window, accept the settings related to profile creation and press OK. You can view the created profile using the *Create Profile View* option in the *Profile* menu, although this is not actually needed.

Now, to insert the label, open the *Alignments* menu, select *Add Alignment Labels* and then *Add/Edit Station Labels*, and click on the alignment (Figure D20) to open the window shown in Figure D36. In this window, to create the label at the stations included in the profile, set the *Type* menu to *Profile Geometry Points* and then click on the *Add* button to open a window where you must select the profile (Figure D37). Doing this will add the parameter to the lower pane, as shown in Figure D36.

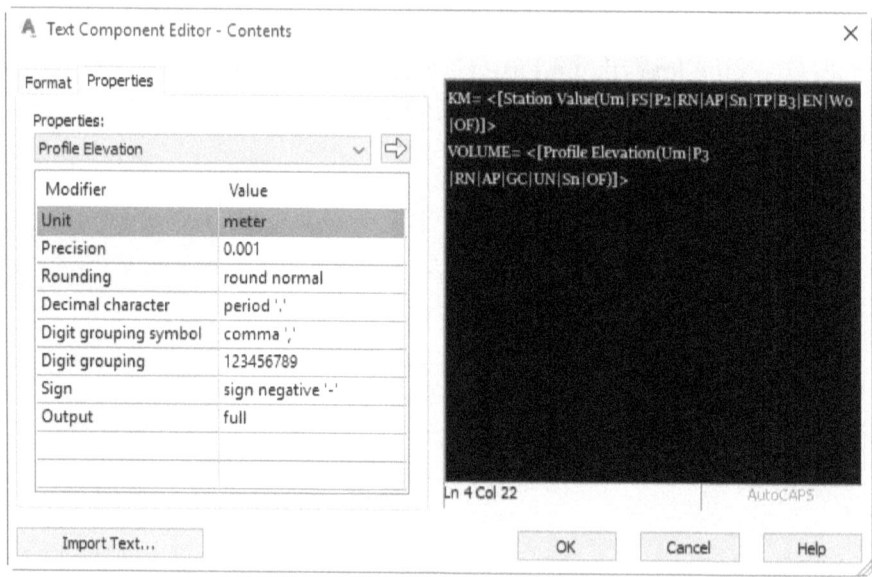

Figure D39

Now, to configure how parameters are displayed, you must click on the small icon under the *Style* column (see Figure D38). In the opened window, click on the arrow button and choose *Edit Current Selection* to open the window shown in Figure D33. To edit the parameters displayed in the label, click on the field in front of *Contents* to open another window (Figure D39).

In this window, delete the contents of the right pane and type *KM=* in its place. Then open the *Properties* menu on the left pane and select *Station Value* and press the arrow button to insert it in the right pane in front of *KM=*.

If you want to the volume of each section to appear below the station label, type *Volume=* in the next line, find *Profile Elevation* in the *Properties* menu and use the arrow button to insert it at the end (this only works if you insert the volume values instead of elevation values in the second column of the excel file). Press OK to close the windows. Ultimately, the station labels must appear as shown in Figure D40.

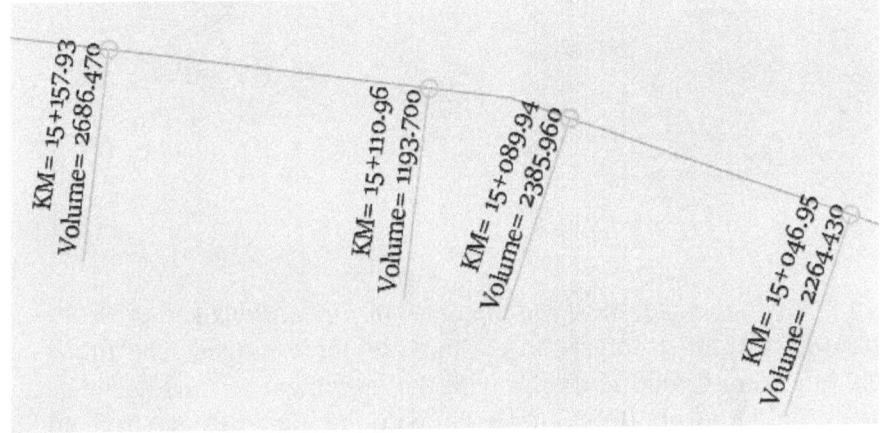

Figure D40

Note: If you want the labels to appear without *KM=* and *Volume=*, simply do not type them in the label editing pane.

49- How to create station labels at 5-meter intervals over one segment of an alignment and at 20-meter intervals over another segment?

For this exercise, open the file al-1.dwg from the folder Project File. This file is a drawing of a 1599.13 meters long route. The goal is to create station labels at 5meter intervals for the segment between 00+600 and 00+900 and at 20meter intervals for the rest of the route. To do this, open the *Alignments* menu, select *Add Alignment Labels*, and then *Add/Edit Station Labels* (Figure D41) to open the window shown in Figure D42.

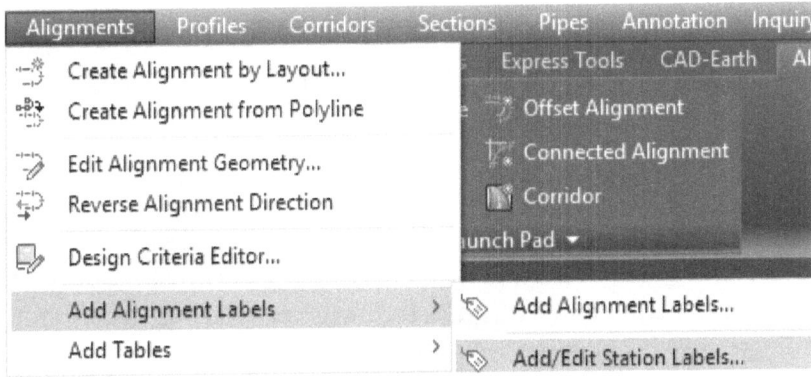

Figure D41

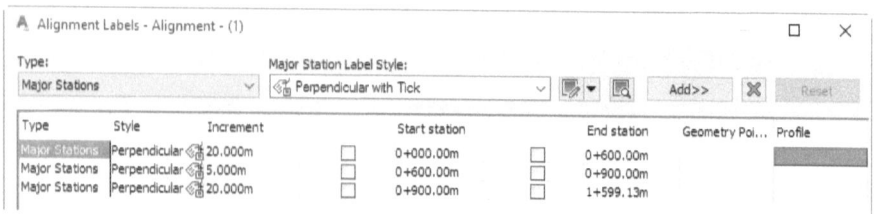

Figure D42

In this window, set the *Type* menu to *Major Stations* and press the *add* button three times to create three lines for three ranges. The first range should be from 0+000 to 0+600 with the increment set to 20. The second range should be from 0+600 to 0+900 with the increment set to 5, and the last range should be from 0+900 to 1+599.13, which is the end of the route, with the increment set to 20 again (see Figure D42). When done, press OK to close the window.

50- How to number the points of intersection (PIs)?

In many CIVIL3D project, it is imperative to number and index the PIs of the route (alignment). As you know, CIVIL3D does not automatically number the PIs (normally, these points are simply named PI without any number). To number the PIs manually, you have to define station equations for PIs, give each PI a name while doing so, and make the station equations appear in labels.

Figure D43

For this exercise, open the file <u>AL-PI.dwg</u> from the folder <u>Project File</u>. This file contains a alignment consisting of ten PIs (start and end points included). The goal is to number the PI from 1 to 10 using labels. Click on the alignment to select it, open the *Alignment* tab in the ribbon and select *Alignment Properties* (Figure D43).

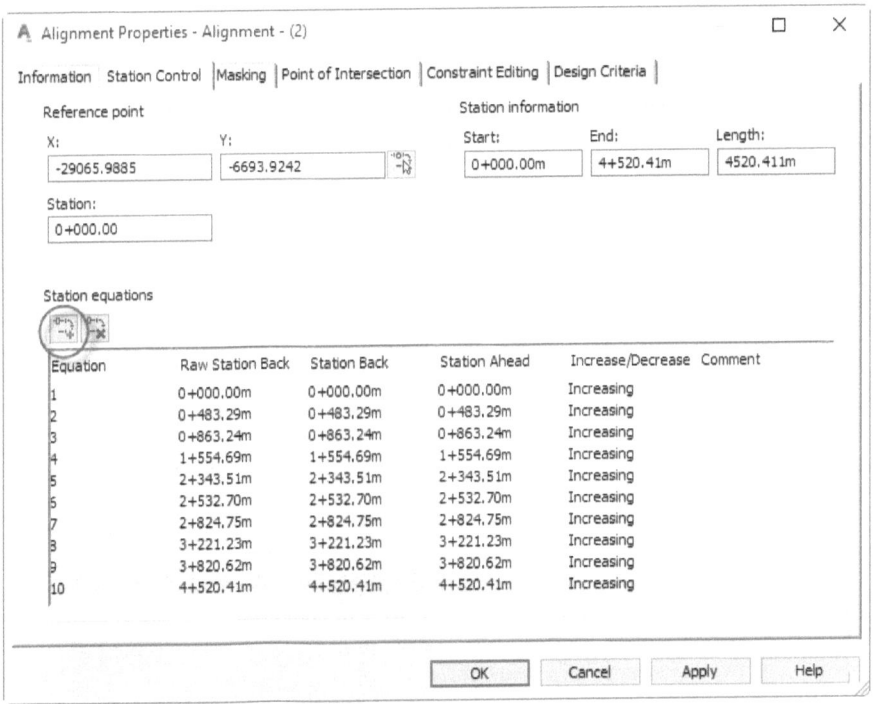

Figure D44

This opens the *Alignment Properties* window displayed in Figure D44. In this window, go to the *Station Equation* tab, press the *Add Station Equation* button (marked in the figure) and click on the PIs of the alignment until all ten of them are added to the list below. If you like to give the PIs a name rather than a number, you can click on the PI name in the equation column to change the name. Once finished, press OK to close the window.

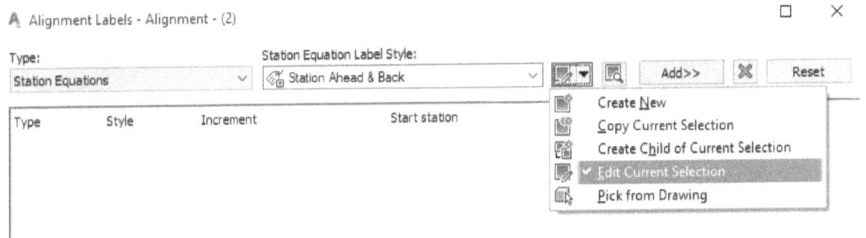

Figure D45

Now that the station equations are defined, you have to create and set up labels. For this purpose, open the *Alignments* menu, select *Add Alignment Labels*, and then *Add/Edit Station Labels* (Figure D41) to open the window shown in Figure D45.

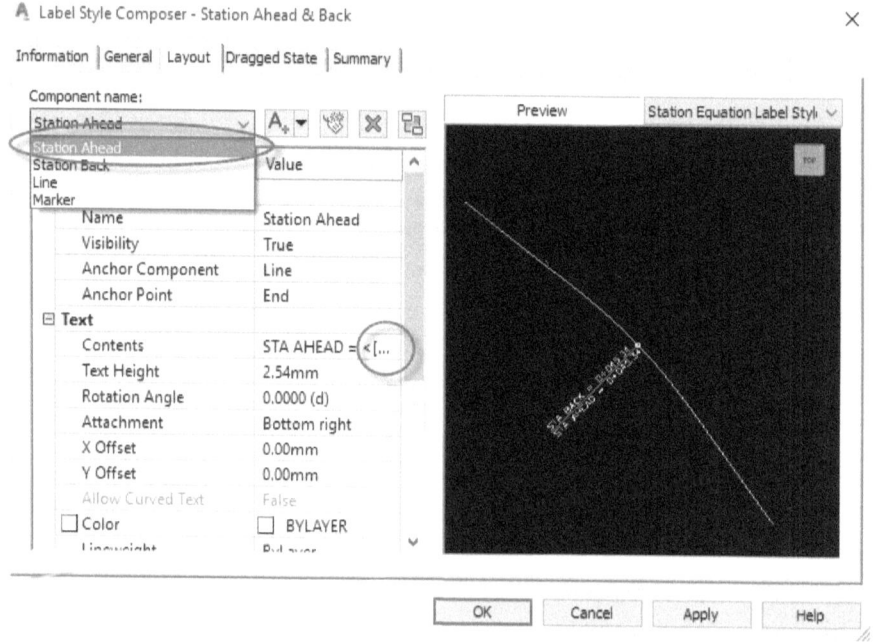

Figure D46

In this window, set the *Type* menu to *Station Equations* and then press the button to the right of Station Ahead & Back (see the figure) and select *Edit Current Selection* (alternatively, you can select *Create New* to create a new label).

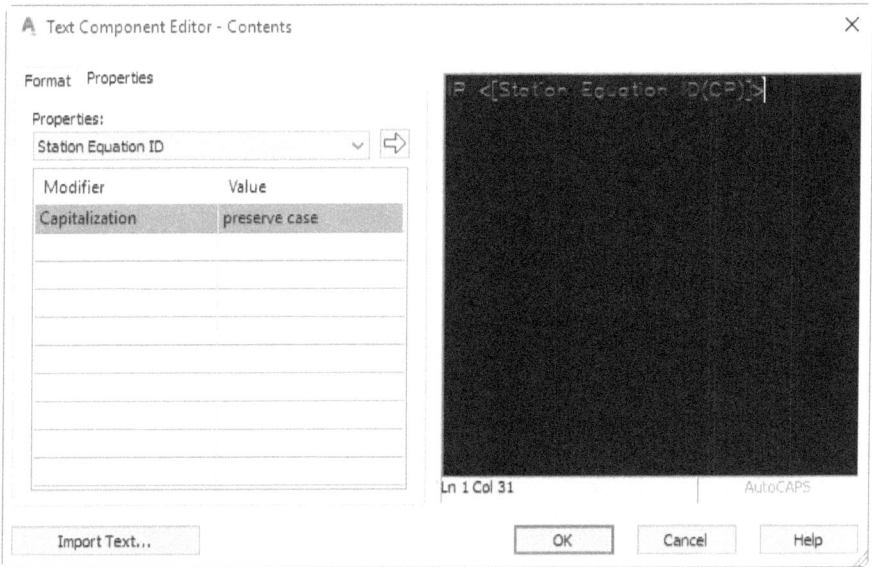

Figure D47

In the opened window (Figure D46), delete either *Station Ahead* or *Station Back* and then select the other one and click on the field in front of *Contents* to configure the numbering labels.

As shown in Figure D47, delete the contents of the right pane, type *PI* (if you want the word PI to appear before the number), select *Station Equation ID* from the *Properties* menu and press the arrow button to add it to the right pane.

If you want the station value to also appear in the label, you must select the *Station Value* from the *Properties* menu and add it to the right pane. Finally, press OK to close the window. After arriving at the *Alignment Labels* window (Figure D45), remember to add the label to the list using the *Add* button.

51- How to change the reference point of an alignment?

To change the reference point of an alignment, select the alignment, open the *Alignment* tab in the ribbon and press the *Alignment Properties* button (Figure D43). In the opened window (Figure D44), go to the

Station tab and change the reference point in the box named *Reference point.*

52- How to edit the horizontal parameters of an alignment (e.g. curve radius, spiral length, etc.)?

There are two ways to edit the horizontal parameters of an alignment. But first you must open the *Alignments* menu and select *Edit Alignment Geometry* to open the toolbar shown in Figure D48.

Pressing Button No.1 in Figure D48 and then clicking on any component of the route will open a window that allows you to make desired adjustments in that component.

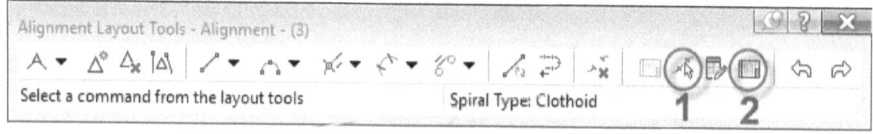

Figure D48

By pressing Button No.2 in that toolbar, you gain access to a list of all curve parameters, where you can click on any parameter and edit it as needed.

53- How to define different design speeds for different segments of an alignment?

When using the *Layout* feature and software standards to design a route, you can define different design speeds for different segments. To do this, open the *Alignments* menu and select *Alignment Properties*. In the opened window, go to the *Design Criteria* tab and use the button shown in Figure D49 to add the start station of the segments that ought to have different design speeds and then specify the design speeds in the third column.

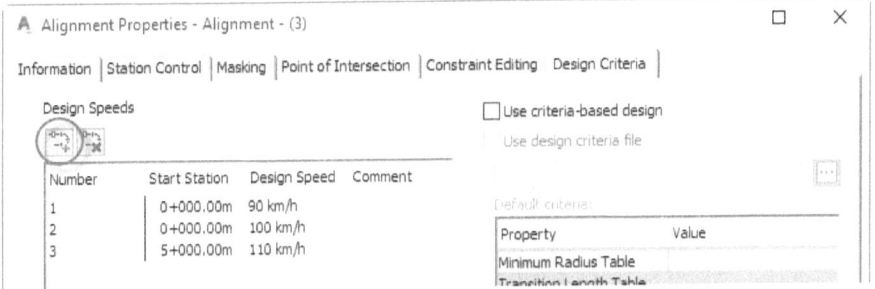

Figure D49

54- How to make offset profiles appear over the longitudinal profile with a different color than the axis profile?

One issue of great importance for the design of project line is the state of the natural ground surface at different cross sections over the longitudinal profile. For example, consider the cross section portrayed in Figure E1. Given only the axis profile, one may think that the project line of 1359.30 can be achieved with just a 30cm excavation, but after seeing the cross section, it becomes clear that the right side of the road would then need to be filled by as much as 3.60 meters. A simple solution to prevent such misconceptions is to make longitudinal profiles include not only the axis profile but also the offset profile. Depending on the nature of the project, designers often prepare longitudinal profiles at short regular distances from the main axis (e.g. at 1 meter intervals). For a 7 meters wide route, for example, longitudinal profiles may be drawn at -3.5, -2, -1, 1, 2, and 3.5 meter distances from the axis (along with the axis, they make a total of 7 profiles). Also note that for easier identification, it is common to color the profiles on the left side of the axis differently than those on the right.

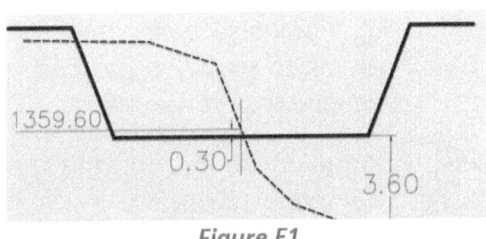

Figure E1

For this exercise, open the file <u>Profile-1.dwg</u> in the folder <u>Project File</u>. This file contains the drawing of a route (alignment) and the topography of the region. The goal is to produce the longitudinal profile for the axis as well as the 1, 2, and 3.5m bands on its left and right side. Before creating the profiles, you need to create three profile styles for assigning different colors to the left, axis, and right profiles. To do this, open the *TOOLSPACE* window, go to the *Settings* tab, right-click on the *Profile Style*, and select *New* to open the window shown in Figure E3.

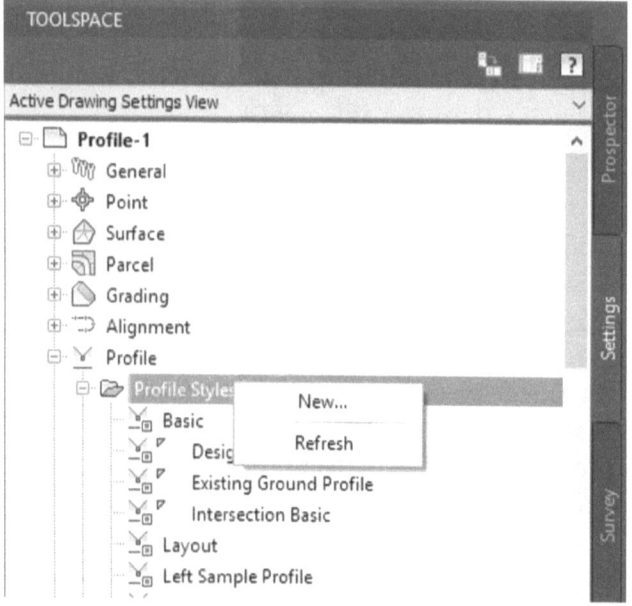

Figure E2

In this window, first go to the *Information* tab and type a name for the new style, then go to the *Display* tab and look for a box named *Component Display*. Find the component type named *Line* and click on the same row in the *Color* column to specify a color for this component type (Figure E3). Press OK to create the profile style.

Follow the above instruction to create three profile styles named Axe, Left, and Right, and assign a color to each one.

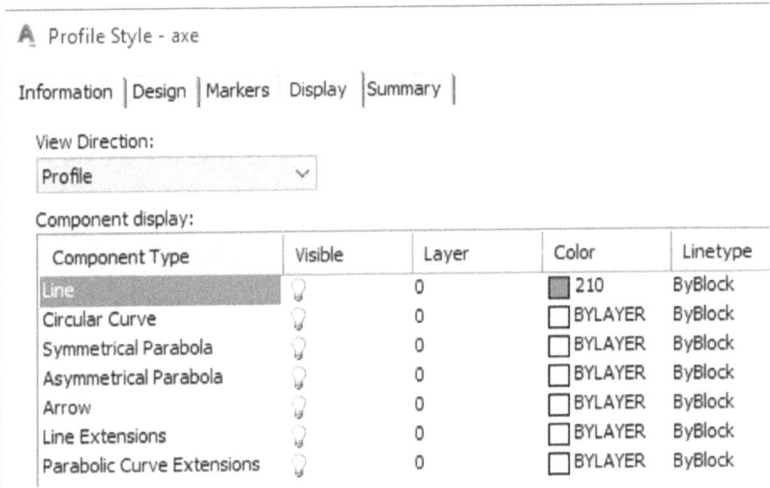

Figure E3

Then, open the *Profile* menu and select *Create Profile From Surface.* This will open the window shown in Figure E4.

A Profile Style - axe

Information | Design | Markers | Display | Summary |

View Direction:

Profile

Component display:

Component Type	Visible	Layer	Color	Linetype
Line		0	■ 210	ByBlock
Circular Curve		0	☐ BYLAYER	ByBlock
Symmetrical Parabola		0	☐ BYLAYER	ByBlock
Asymmetrical Parabola		0	☐ BYLAYER	ByBlock
Arrow		0	☐ BYLAYER	ByBlock
Line Extensions		0	☐ BYLAYER	ByBlock
Parabolic Curve Extensions		0	☐ BYLAYER	ByBlock

Figure E3

In this window, open the *Alignment* menu and choose the alignment of interest and also select the surface of interest in the *Select Surfaces* menu. Then press the *add* button to add the axis profile. To create the profile at the offsets of interest, uncheck the box of *Sample offsets* and type 1 in the field below and press the *Add* button. Repeat this process by typing 2, 3.5, -1, -2, and -3.5 (one at a time) in the field and pressing the *add* button. In the end, these profiles should appear in the profile list as shown in Figure E4. For the left, right, and axis profiles to appear in different colors, you need to assign the styles you created in the previous step to these profiles. This can be done by clicking under the *Style* column (see Figure E4) and selecting the name of the appropriate created profile (we named them *Left*, *Right*, and *Axe*). In the end, click on the *Draw in profile view* button to open a window, where you must press *Create Profile View* button and then click on a point on the drawing to draw the profile.

55- How to display the elevation of longitudinal profile at each offset in separate bands under the profile?

For this exercise, open the file <u>Profile-2.dwg</u> in the folder <u>Project File</u>. This file contains the topography of an area, a route (alignment) and its longitudinal profile, with no bands displayed below the profile. The profile provided in this file actually consists of seven minor profiles each dedicated to one specific offset. The goal is to insert seven bands each showing the elevation of one of these minor profiles. To do this, click on the profile to select it and then press the *Profile View properties* button on the ribbon (Figure E5) to open the window displayed in Figure E6.

Figure E5

In this window, go to the *Bands* tab and set the *Band type* menu to *Profile Data* and the *Select band style* menu to *Elevation*. Then use the *add* button to add seven bands to the list (because here we have seven minor profiles). Specify the minor profile to be assigned to each band in the Profile 1 column. When done, press OK to close the window.

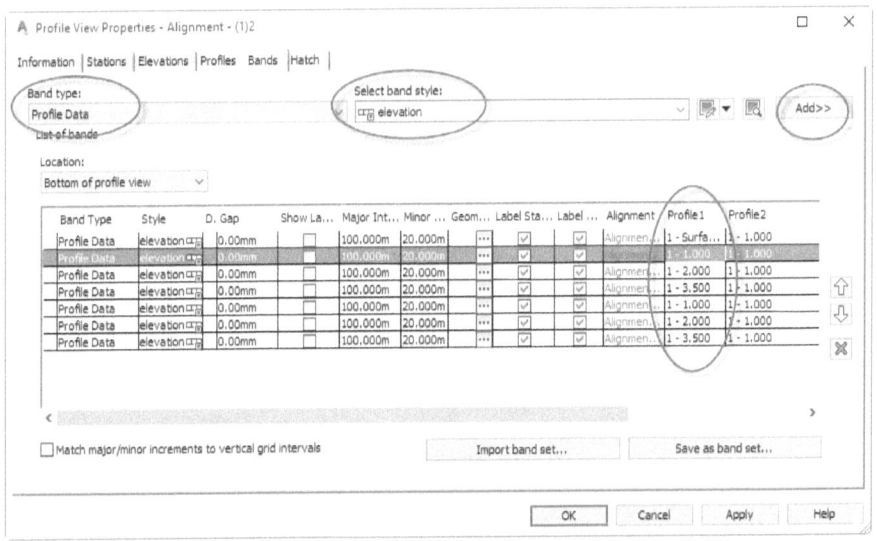

Figure E6

56- How to customize the label of the slope between two longitudinal profiles?

Sometimes, it is necessary to change the longitudinal slope displayed in the drawing without actually altering the geometry. For example, suppose that we need the slope to be displayed 1% higher than its actual value (16% instead of 15% or -6% instead of -5%). This can be done by creating a conditional expression.

For this exercise, open the file Profile-3.dwg in the folder <u>Project File</u>. This file contains the topography of an area, a route (alignment) and its longitudinal profile. The goal is to apply the described customization (increasing the displayed slope by 1%) on the PIs of the longitudinal profile.

Figure E7

Open the *Settings* tab in the *TOOLSPACE* window. In this tab, expand the *Profile* branch and then *Label Styles* and *Line* sub-branches, right-click on Expressions and select *New* (Figure E7). This opens the *New Expression* window shown in Figure E8.

In this window, type a name for the expression to be created, then use the buttons below to enter the following conditional expression in the *Expression* box:

IF({Tangent Grade}>0,{Tangent Grade}+0.01,{Tangent Grade}-0.01)

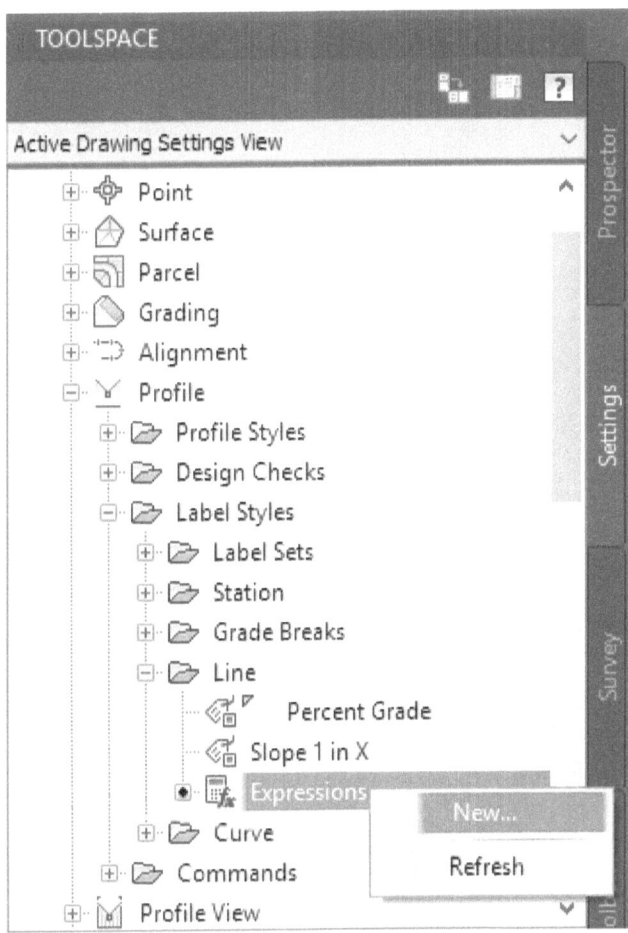

Figure E8

This conditional command states that the longitudinal slope should be increased by 1 percent if it is positive and should be decreased by 1 percent otherwise. Note that since this expression works with the slope as a percentage, the *Format result as* menu should be set to *Percent*. When finished, press OK to close the window.

Now you have to create a *Label Style* for using this expression. To do this, return to the *Line* sub-branch in the *Settings* tab (Figure E7), but this time right-click on *Line* and select New. Figure E9.

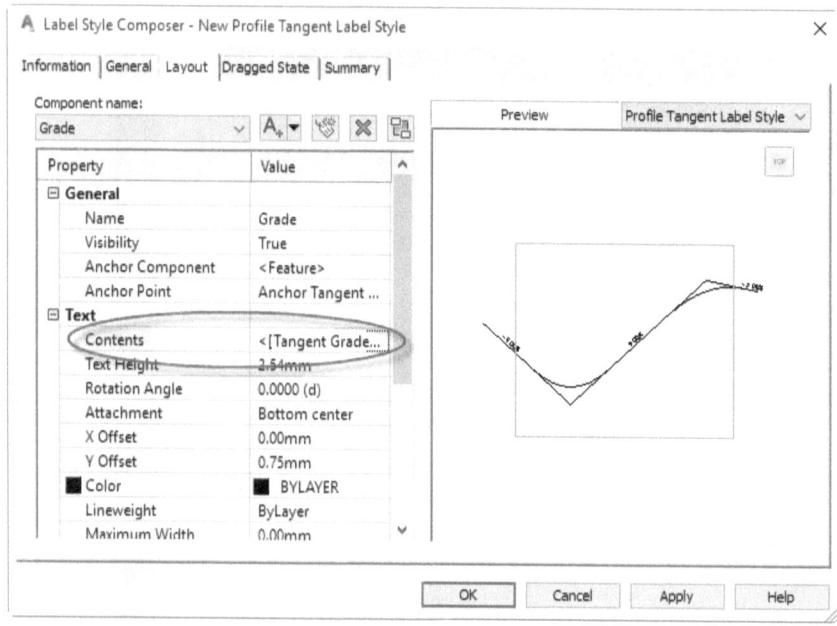

Figure E9

In the opened window (Figure E9), type a name for the style to be created in the *Information* tab, and then go to the *Layout* tab.

In the *Layout* tab, expand the *Text* branch and click on the field in front of *Contents* to open the window shown in Figure E10.

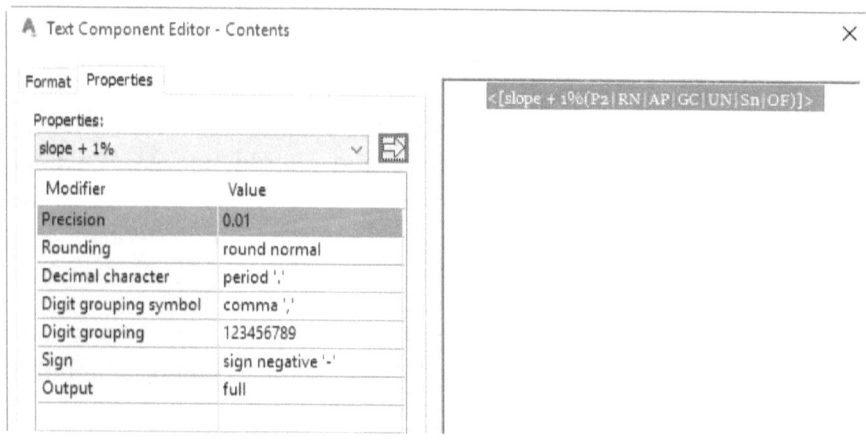

Figure E10

In this window, delete the contents of the right pane, find the expression you created in the *Properties* menu, set the *Precision* to two decimals and click on the arrow button to add the expression to the right pane.

To finish creating a label style, close this window and the previous one by pressing OK.

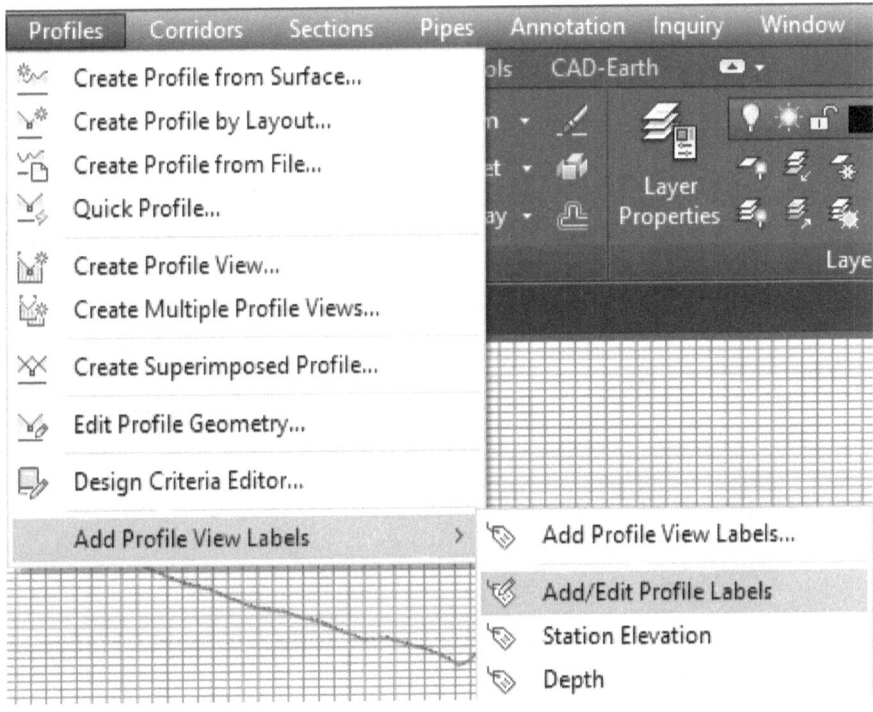

Figure E11

Now that the label style is created, you have to apply it to the profiles. To do this, open the *Profile* menu and select *Add Profile View Labels* and then *Add / Edit Profile Labels* (Figure E11). At this stage, the software will ask you to click on the profile to which label style must be applied. After doing so, you will be directed to the window shown in Figure E12.

In the window, set the *Type* menu to *Line*, then open the *Profile Tangent Label Style* menu and select the label style you created in this previous step (we named it SLOPE-PERCENT-1). Press the add button to add the style to the pane below. In this pane, you can specify the segment to which style must be applied. Once finished, press OK to apply the changes on labels.

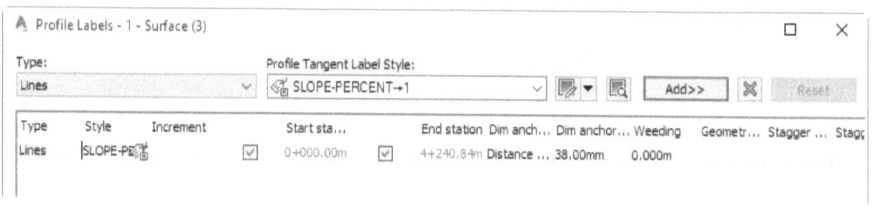

Figure E12

In Figure E13, the labels displayed below the line show the real slope and those above the line show the customized value. As you can see, the upper labels display the slope values of -4.63% and 8.27% while the actual values are -3.63% and 7.27%.

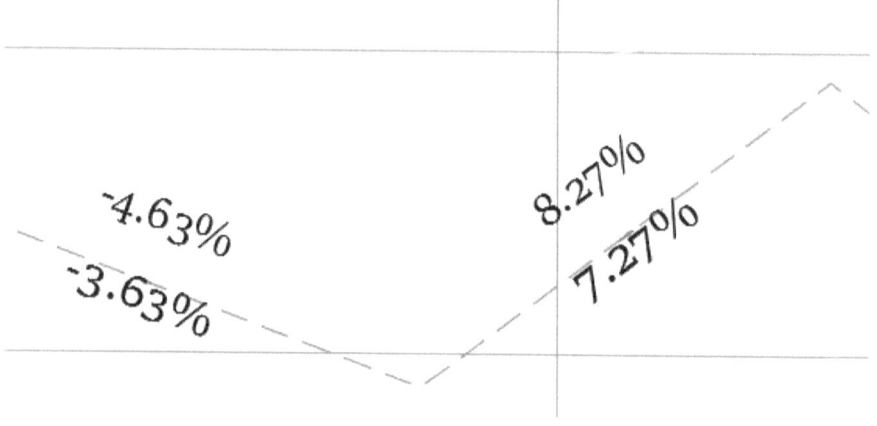

Figure E13

57- How to display the horizontal parameters of an alignment as a band under the profile?

One of the common needs of CIVIL3D users is to insert the information such as curve length, curve radius, arc deviation, left turns or right turns, etc. as a band below the longitudinal profile (Figure E14).

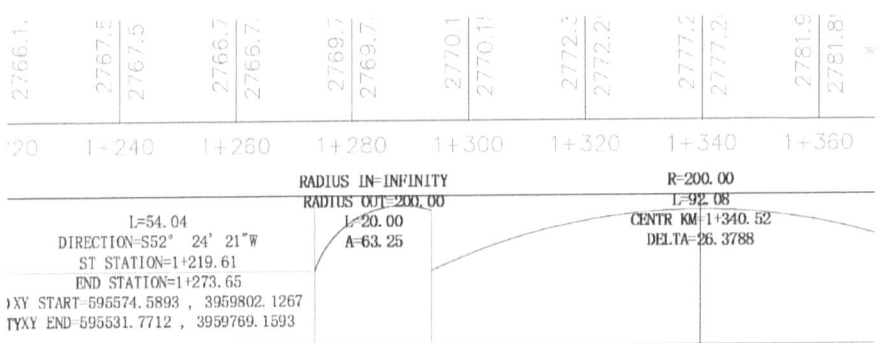

Figure E14

This information can be divided into four groups: those related to straight alignments (tangent), those related to curves, those related to spirals, and those related to intersection points.

- For straight alignments, this information include length, orientation, coordinates of start and end points, etc.

- For curve, this information include curve length, chord length, central angle, coordinates of start and end points, etc.

- For spirals, this information include arc length, parameters A and K, entrance and exit radius, etc.

- For points of intersection, this information include coordinates, number, station, etc.

Depending on the type of project, the amount of details needed, and the preferred way of presenting information, you can create separate bands for each of the aforementioned types of information, or put all of them in one band. Here, we explain how to create separate bands for each type of information, but with some practice, you will be able to make the software display the information as you like.

For this exercise, open the file Profile-Band-1 in the folder Project File. This file contains a surface, a route (alignment) with horizontal curves and intersection points, and a longitudinal profile. The goal is to insert several bands under the longitudinal profile, each dedicated to one of the aforementioned information types.

Straight alignment (tangent)

To create a band, you need to first create a style for that band. To do this, open the *Settings* tab in the *TOOLSPACE* window, and expand the *Profile View* branch and then *Band Styles* and Horizontal *Geometry* sub-branches (Figure E15). Right click on *Horizontal Geometry* and select *New* to create a new style for this type of bands.

This opens the window shown in Figure E16.

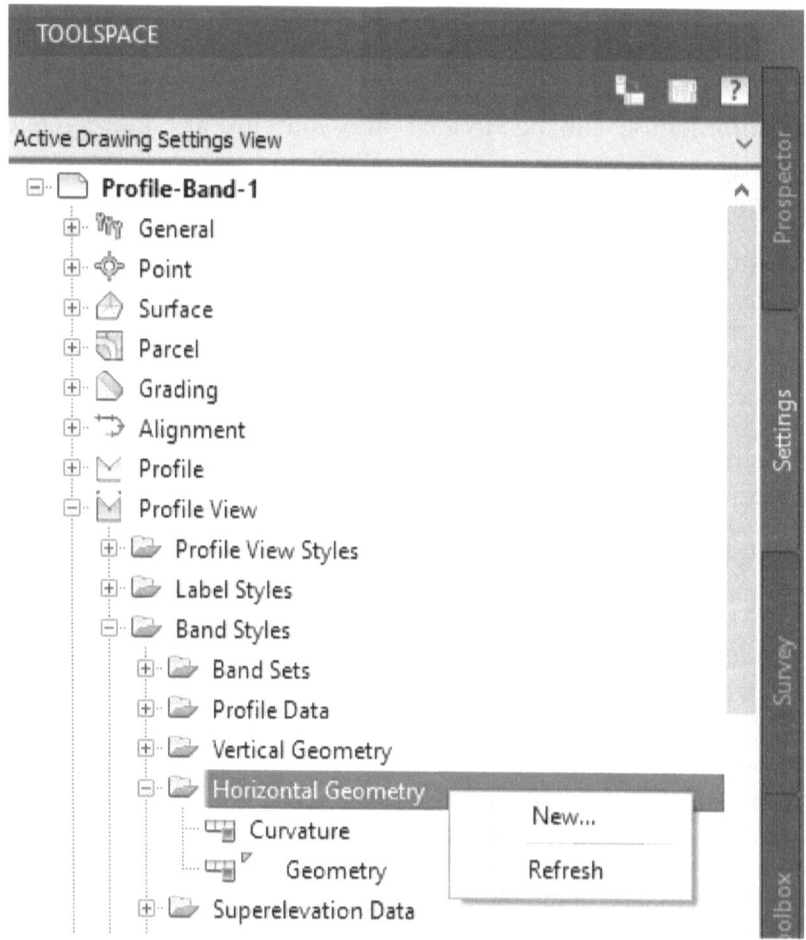

Figure E15

In this window, go to the *Information* tab and type a name for the band style to be created (here, we named the band style *Tangent*).

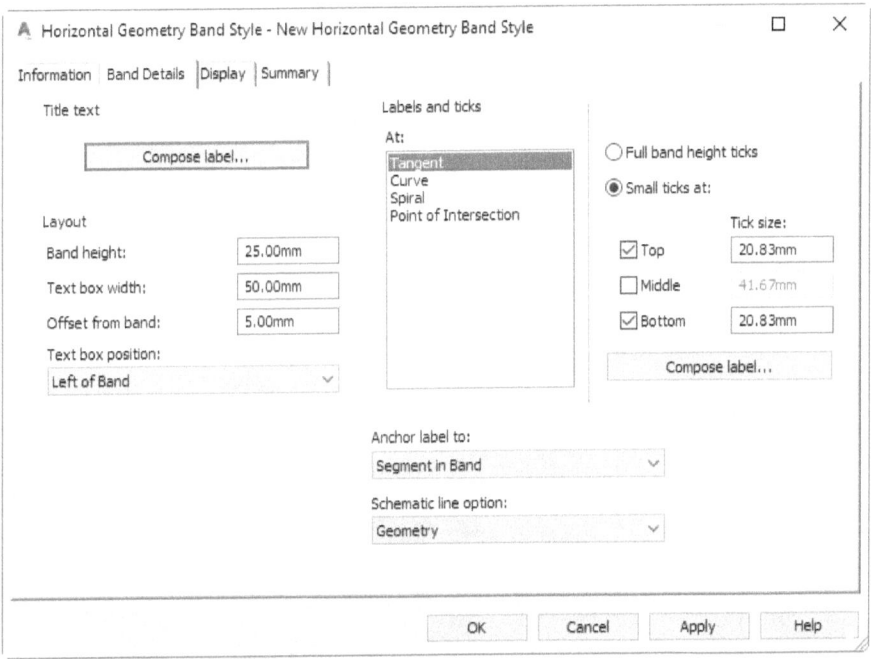

Figure E16

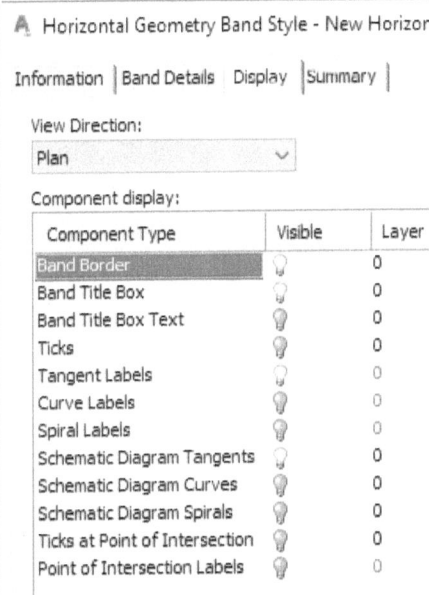

Figure E17

Every band style to be defined here will cover all four types of information, but here we want only the information related to tangent to appear. There are two ways to hide the other three information types. The easiest way is to go to the *Display* tab and turn off the layers related to those information types. As shown in Figure E17, you must turn off every layer except *Tangent Labels* and *Schematic Diagram Tangents*, which are related to tangent, and *Band Border* and *Band Title Box*, which are related to the illustration of band title and borderlines.

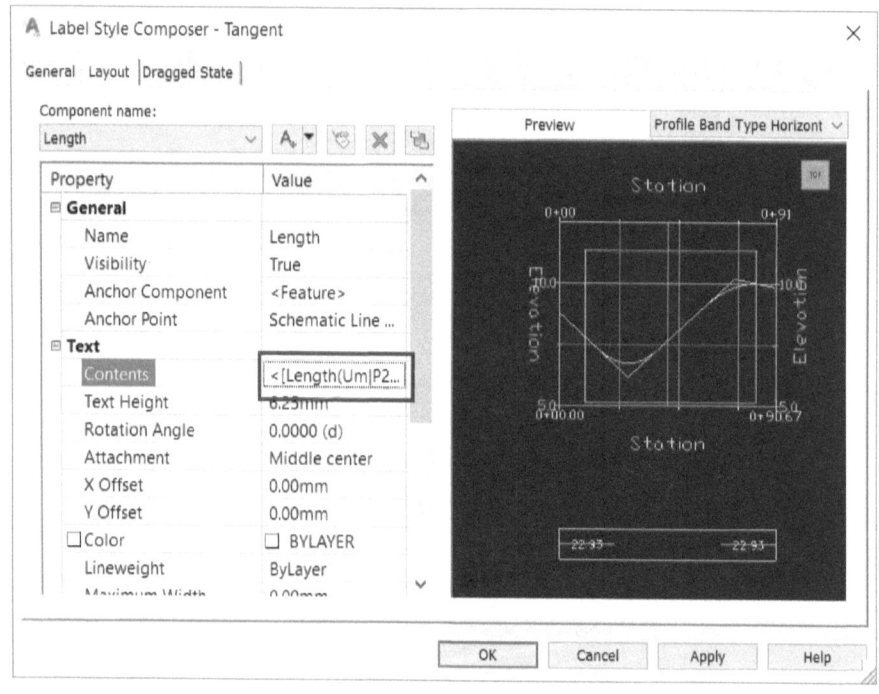

Figure E18

The second method is to go to the *Band Details* tab, click on one of the other information types (*Curve*, *Spiral*, or *Point of Intersection*), press the *Compose Label* button (on the right) and remove the related content in the opened window. After repeating the above procedure for all unwanted information types (here, *Curve*, *Spiral*, or *Point of Intersection*), select *Tangent* and press the *Compose Label* button to open the window shown in Figure E18.

In this window, expand the *Text* branch in the *Property* box and click on the field in front of *Contents* to open the content editing window shown in Figure E19.

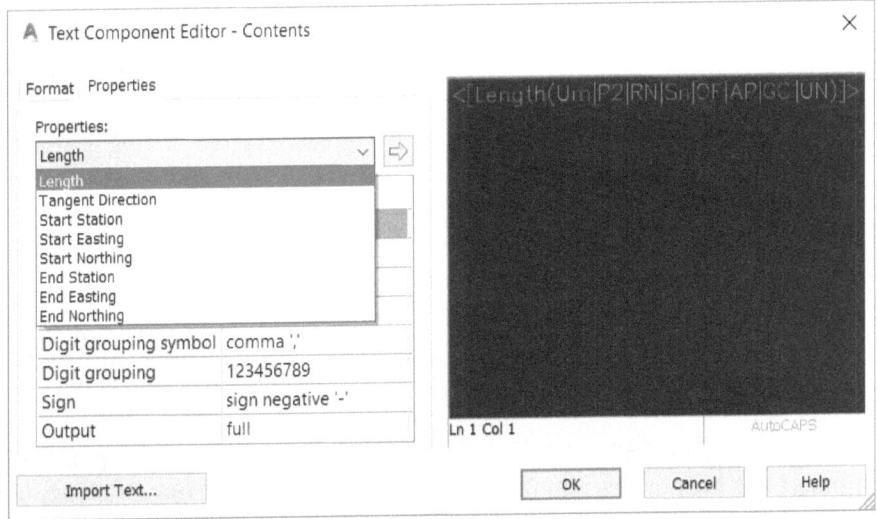

Figure E19

In this window, delete the contents of the right pane, select the property you want displayed from the *Properties* menu, adjust the settings below, and click on the arrow button to insert it to the right. Remember that you can add a title or initials for the property to be inserted by just typing in the right pane.

For example, for the length of a straight alignment, first type *L*= and then insert *Length* and then go to the next line and repeat the process for another property.

Other properties that can be inserted for a straight alignment include:

- *Length Direction* : displays the orientation of an alignment

- *Start Station* : displays the start stations of an alignment

- *End station* : displays the end station of an alignment

- *Start Easting* and *Start Northing*: displays the coordinates of the start point of a straight alignment

- *End Easting* and *End Northing* : displays the coordinates of the end point of a straight alignment

Curve

To insert the band of information related to curves, follow the path shown in Figure E15 to reach at the *Horizontal Geometry* sub-branch in the *Settings* tab. Right click on *Horizontal Geometry* and select *New* to create a new style. Use one of the methods described in the previous subsection to turn off all information types except *Curve*. Go to the *Information* tab and type a name for the band style. As before, select the *Curve* option and press the *Compose label* button to open the window shown in Figure E18. In this window, click on the field in front of *Contents* to open the window shown in Figure E20.

As explained earlier, first delete the contents of the right pane, and then insert the properties you need displayed using the *Properties* menu.

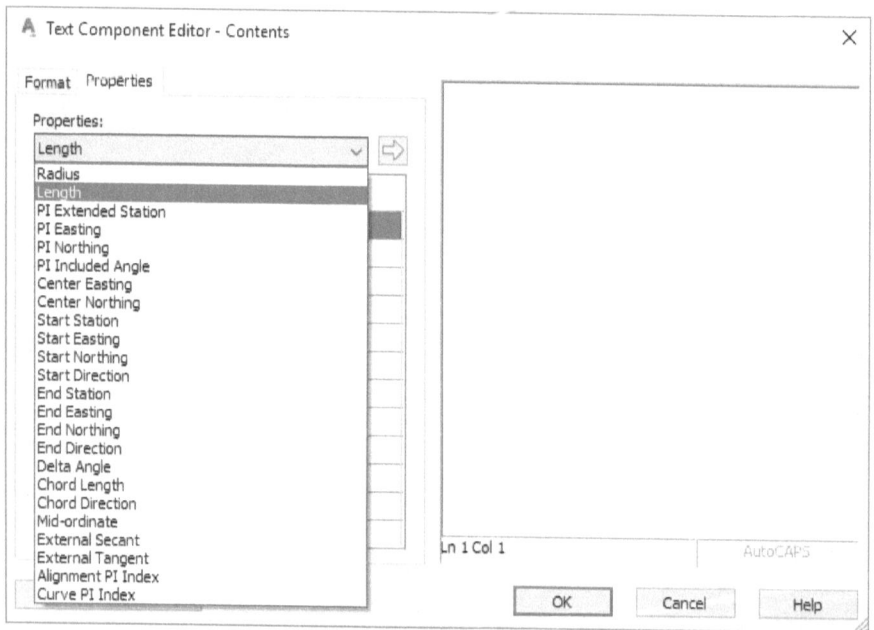

Figure E20

The properties that can be inserted for a curve include:

- *Length* : displays the curve length

- *Radius* : displays the curve radius

- *PI extended Station* : displays the station (chainage) of the curve's PI

- *PI Easting* and *PI Northing* : displays the coordinates of the curve's PI

- *PI Included Angle* : displays ....

- *Center Easting* and *Center Northing* : displays the coordinates of the curve's center

- *Start Station* : displays the station (chainage) of the curve's start point

- *Start Easting* and *Start Northing* : displays the coordinates of the curve's start point

- *Start Direction* : displays the orientation of the curve's start point

- *End Station* : displays the station (chainage) of the curve's end point

- *End Easting* and *End Northing* : displays the coordinates of the curve's end point

- *End Direction* : displays the orientation of the curve's end point

- *Delta Angle* : displays the curve's central angle

- *Chord Length* : displays the length of the curve's chord

- *Chord Direction* : displays the orientation of the curve's chord

- *Mid-ordinate* : displays the mid-ordinate distance

- *External Secant* : displays the external secant

- *External Tangent* : displays the external tangent

- *Alignment PI Index* : displays the index of PI in an alignment

- *Curve PI index* : displays the index of PI in a curve

Band styles for PIs and spirals can also be defined in a similar way.

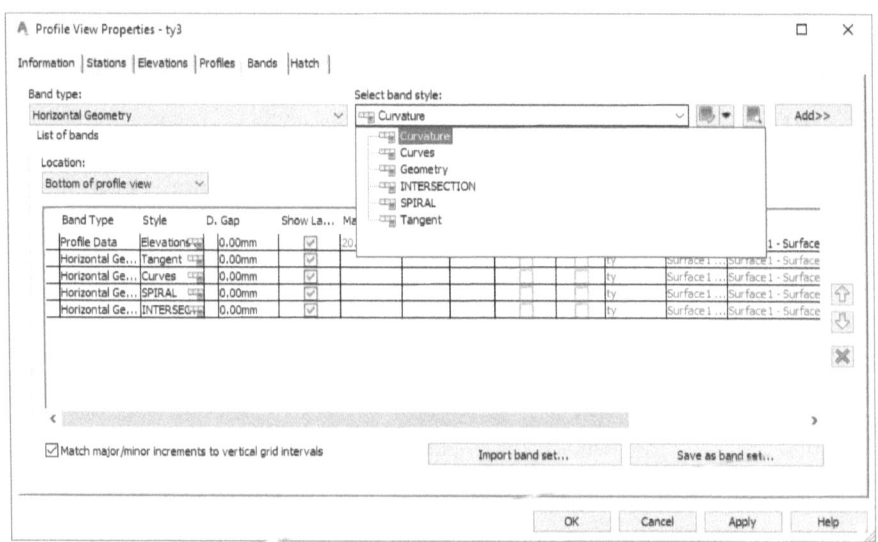

Figure E21

After creating four band styles for the four aforementioned information types, they have to be applied to the profile. For this purpose, select the profile, right-click on it, and select *Profile View* to open the window shown in Figure E21. In this window, go to the *Band* tab and set the *Band type* to *Horizontal Geometry*. Open the *Select band style* menu, and select and add the created band styles one at a time. After pressing Ok, four bands should appear in the drawing as shown in Figure E22.

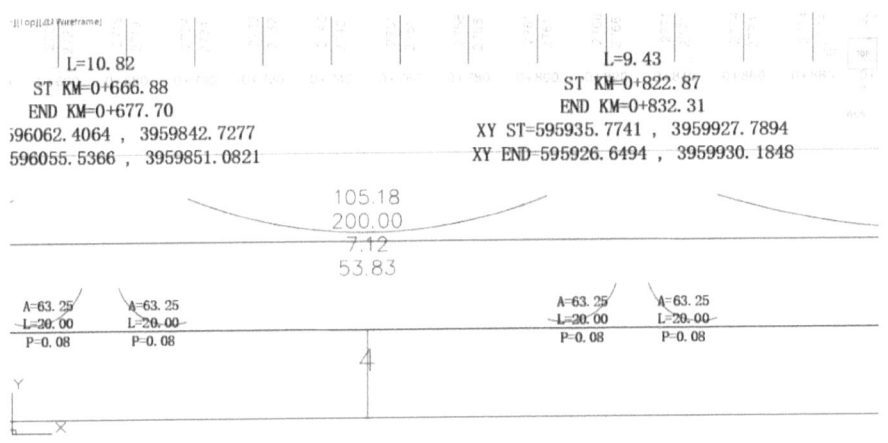

Figure E22

58- How to assign custom colors and boldness to the longitudinal profile of the curved segments of an alignment?

In some projects, it is easier to draw the project lines or design the project structures if you know which segments of a profile are on a curve and which are on a straight line. But one of the weaknesses of CIVIL3D is that it does not display two-dimensional objects and parameters on profiles. However, this problem can be resolved by combining several commands. In Figure E23, for example, the segments of the profile where the route (alignment) is curved are plotted with bold blue lines to give the user better control over the design.

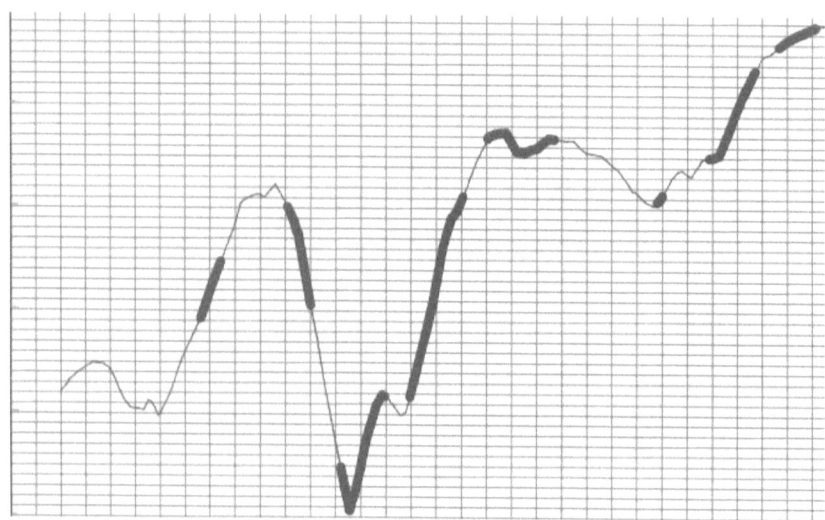

Figure E23

For this exercise, open the file <u>Profile-Horizantal-1</u> from the folder <u>Project File</u>. This file contains a surface, a route (alignment) with horizontal curves and intersection points, and a longitudinal profile. The goal is to apply custom colors and boldness to the longitudinal profile in places where the alignment is curved.

Figure E24

To do this, first select the alignment and copy it over itself. Then, use the *Explode* tool (with the shortcut X) to explode the alignment two times to turn it into arcs (curved lines) and lines (direct lines or tangents). Note that with the first explode, the alignment will only be converted to block

Note: Be careful not to explode both alignments, or in other words, make sure to apply the second explode just on the block that is created after the first explode.

The next step is to turn the arcs into feature lines. For the sake of convenience, select one of the arcs and then right click on it and click on *Select Similar* to select all similar arcs (Figure E24).

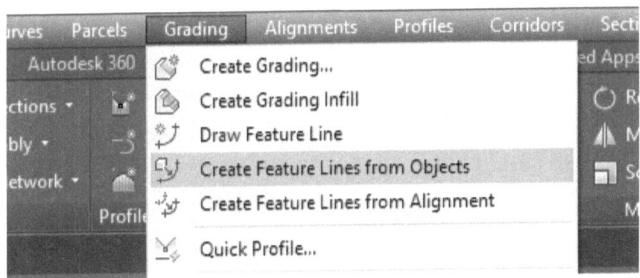

Figure E25

To convert the selected arcs to feature line, open the *Grading* menu and select *Create Feature Lines from Objects* (Figure E25) to open the window shown in Figure E26. At the bottom of this window, check the box of *Assign elevation* option to make the software derive the elevations of the arcs from the surface. If you leave this option unchecked, the elevations will be set to 0 and you will need to define the elevation range of the profile from 0, which leads to a mismatch between the project and its profile.

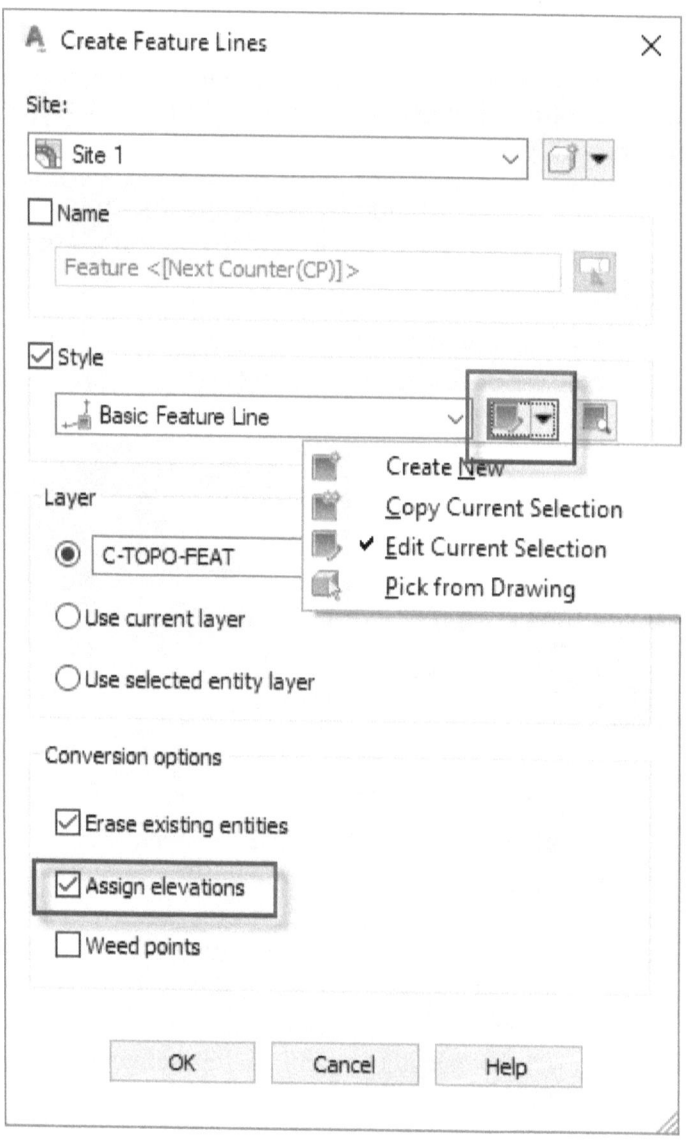

Figure E26

Tick the *Style* option and select one of the styles. Then click on the cascading menu and select *Edit Current Selection* to open the window shown in Figure E27.

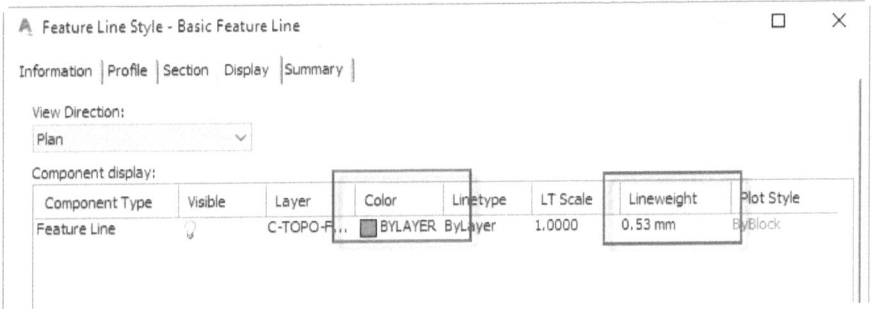

Figure E27

In this window, go to the *Display* tab and specify the line color and boldness in the *Color* and *Line weight* fields. Press OK to close the window and return to the *Create Feature Lines* window (Figure E26).

After closing the *Create Feature Lines* window, you will be directed to the window shown in Figure E28. In this window, you must select the surface from which the elevation of the arcs should be derived. Select the natural ground surface if you want the profile elevation to match the natural surface of the land, or the corridor surface if you want it to match the project line. When done, press OK to close the window. Now that the arcs are converted into feature lines at the elevation of the ground surface, the next step is to make them appear on the profile. To do so, open the *Home* tab in the ribbon and find the *Profile View* button in the *Profile & Section Views* panel. Press this button and then select *Project Object To Profile View* (Figure E29).

Now you have to select all the arc that were converted into feature line.

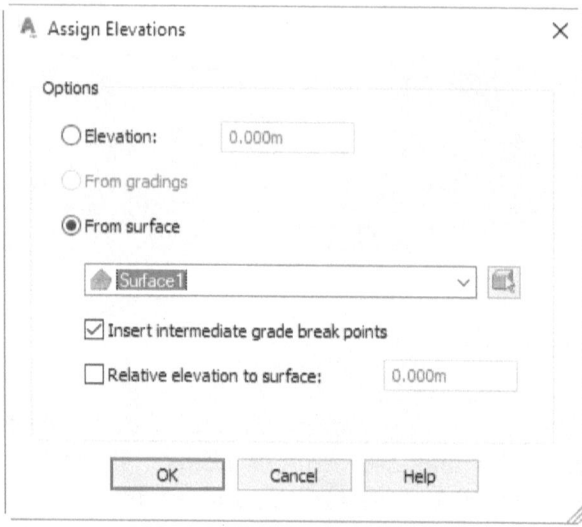

Figure E28

It is easier to first select the feature lines by right clicking and selecting *Select Similar*, and then project them onto the profile. In that case, a message on the command line will ask you to *Select a profile view*. Click on the desired profile to start the operation. After choosing the profile, another window will be displayed for selecting the style. Since you edited and selected the style in advance, there is no need to make any change and you can just press OK to confirm the selected styles.

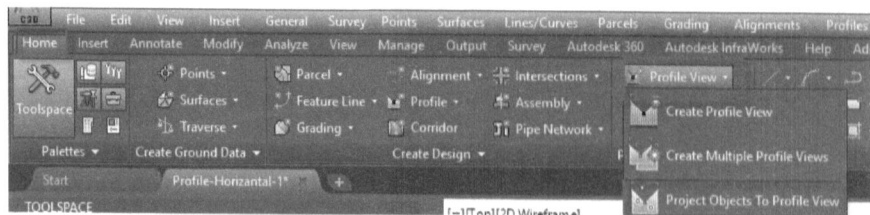

Figure E29

59- How to display the features located along a route on its longitudinal profile?

For this exercise, open the file <u>Profile-Feature-on-Band.dwg</u> in the folder <u>Project File</u>. As shown in Figure E30, the alignment drawn in this file has a box girder bridge and a canal located at stations 0+170 and 0+240. The goal is to mark the position and elevation of these features on the longitudinal profile.

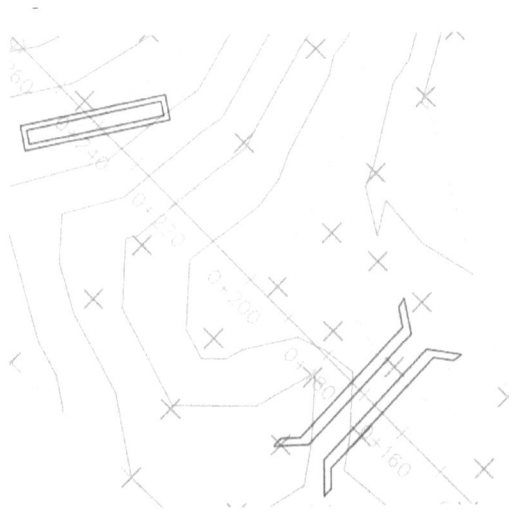

Figure E30

When adding a feature to a longitudinal profile, it is important to remember that the features have their own elevations. The elevation of natural features such as streams must be derived from the natural ground surface. For manmade features such as bridges and canals, the elevation values should be obtained from the design. In this exercise, the elevations are going to be derived from the design.

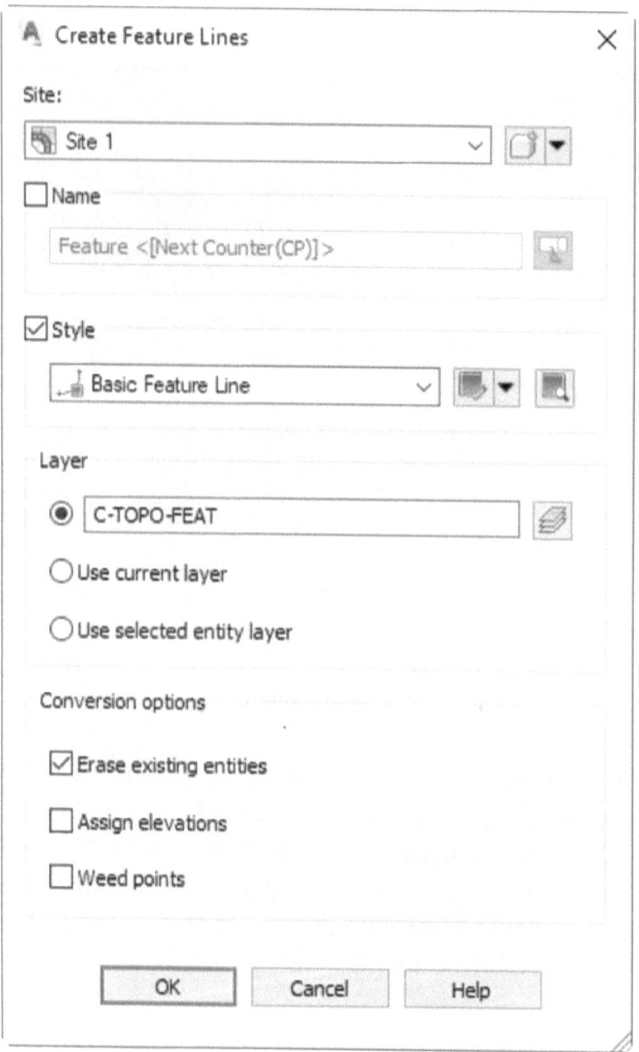

Figure E31

To display these features on the profile, you should first convert them into feature lines. To do this, open the Grading menu and select *Create Feature Lines from Objects* (Figure E25). This opens the window shown in Figure E31. Adjust the settings of this window as you did in previous exercise, but this time leave the *Assign elevation* option unchecked, as features have their own elevations, which must be extracted from the design.

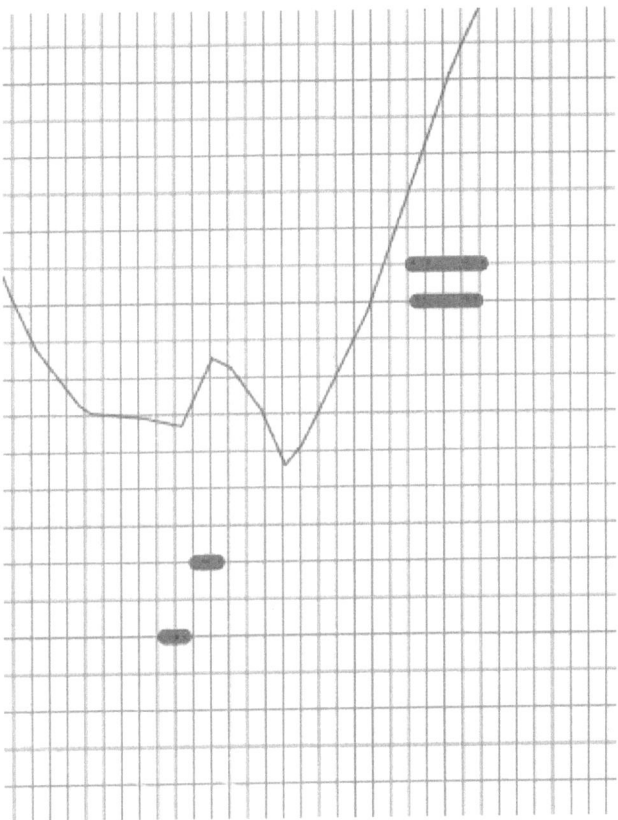

Figure E32

60-How to change the elevation exaggeration of a drawn profile?

This exercise will teach you how to change the elevation exaggeration of a drawn longitudinal profile. For this exercise, open the file Profile-2.dwg. This file contains a longitudinal profile where elevation scale (vertical scale) is 10 times the horizontal scale. The goal is to change this exaggeration ratio to 1 or 20.

To do this, click on the profile, press the *Profile View Properties* button on the ribbon and then select *Profile View Style Edit* (Figure E33).

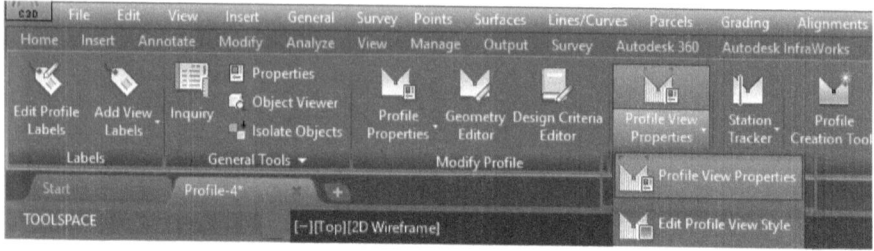

Figure E33

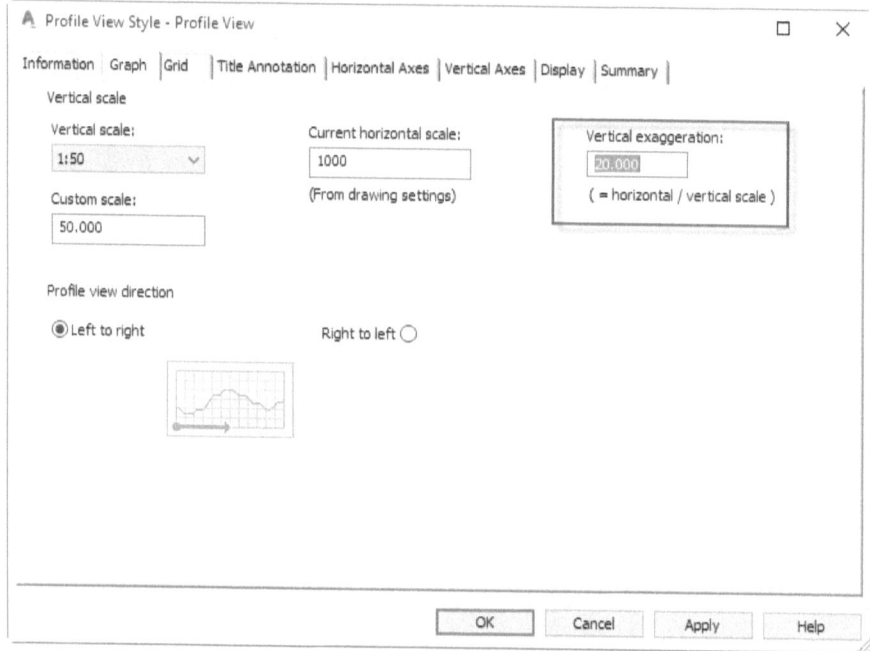

Figure E34

This opens the window displayed in Figure E34. In the *Graph* tab of this window, you can change the value of *Vertical Exaggeration* to 1 to disable exaggeration or to 20 to make the vertical scale 20 times the horizontal scale.

61-How to determine the grade, elevation difference, and distance between two points on a longitudinal profile?

The *Inquiry* tool can be used to determine the grade and elevation difference between two points on a longitudinal profile. Suppose that we want to determine the grade, elevation difference, and distance between two points on the profile of Figure E35. Since the profile may have some vertical exaggeration, it would be difficult to obtain the aforementioned parameters with ordinary AutoCAD tools. An easier way to do this is to use the tool named *Inquiry*. For this purpose, select the profile of interest and then click on the *Inquiry* button on the Ribbon (Figure E36) to open the *Inquiry* window shown in Figure E37.

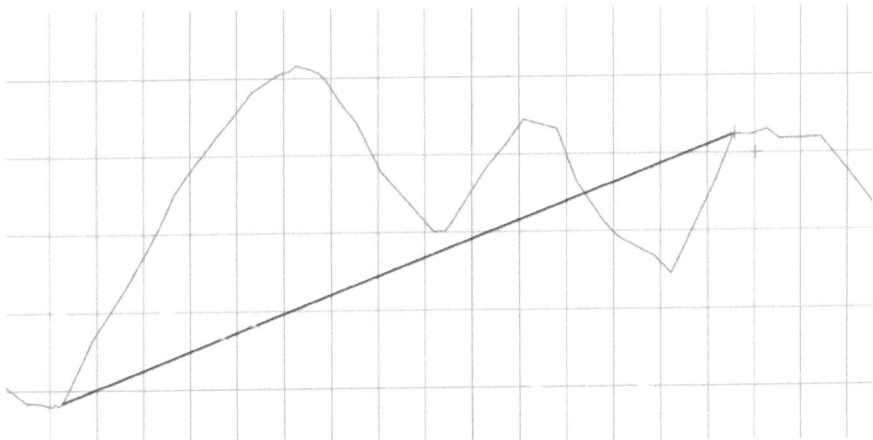

Figure E35

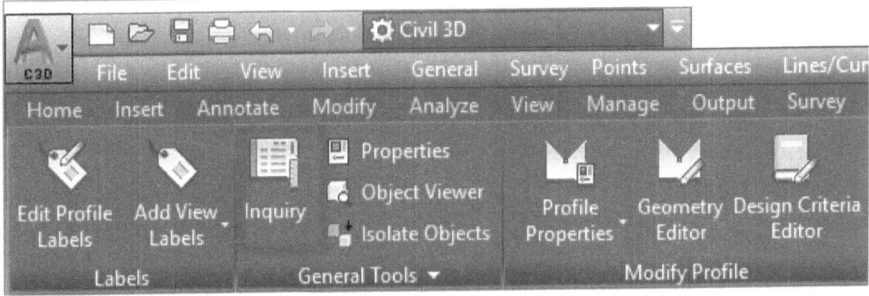

Figure E36

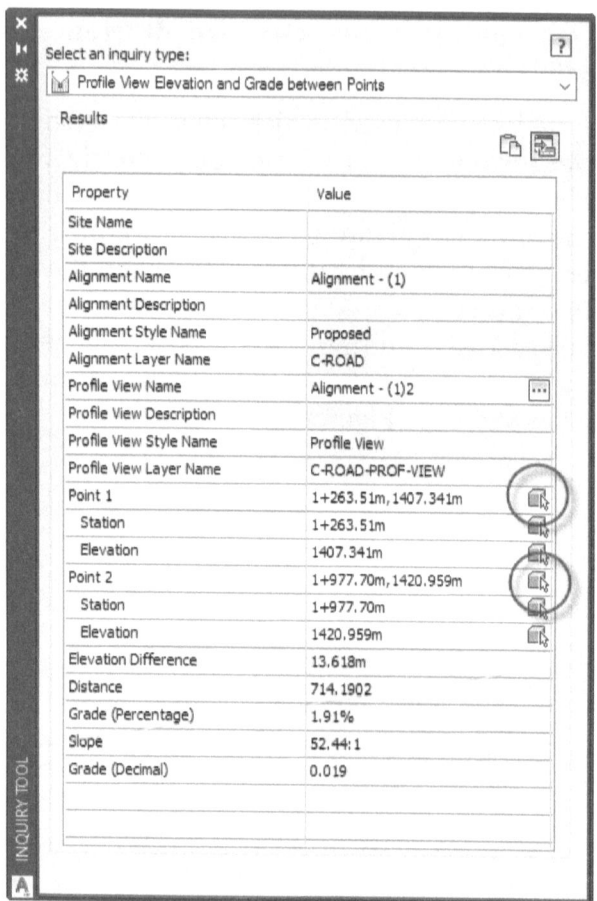

Figure E37

In this window, open the drop-down menu named *Select an inquiry type* and select *Profile View Elevation and Grade between Points* (it can be found under the *Profile View* category)

Now, click on the icons in front of Points 1 and Points 2 boxes and click on the first point and the second point on the profile respectively. The software will then automatically display the elevation difference, distance, and grade (percentage) between the two points. The elevation and station of each point will also be displayed in the boxes of the same name below. Here, you can change Point 1 and Point 2 to easily obtain the aforementioned parameter for any pair of points on the profile. Another feature that designers may find helpful in the determination of grade, distance, and other design parameters is that user can manually change the station and elevation of the selected points and observe the resulting changes.

62-How to draw station and elevation bands on the intersection points of a longitudinal profile?

In the projects involving road and pipeline construction, sometime it is necessary to draw a longitudinal profile with station and elevation information displayed not only at regular intervals (e.g. 20m intervals) but also at critical points or after any grade change in the profile. An example of such profile is shown in Figure E38.

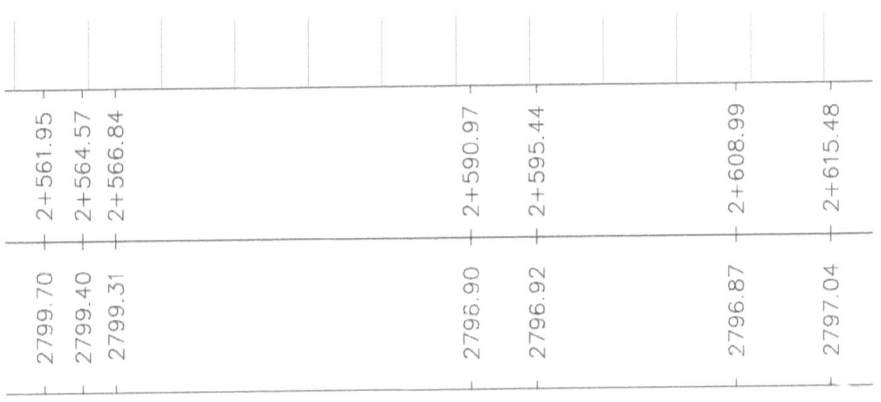

Figure E38

For this exercise, open the file <u>Break Grade-long-profile.dwg</u>, which contains a surface, an alignment and a longitudinal profile. The goal is to draw station and elevation bands under this profile.

To do this, click on the longitudinal profile to select it, then right click on it and select *Profile View Properties* to open the window shown in Figure E39.

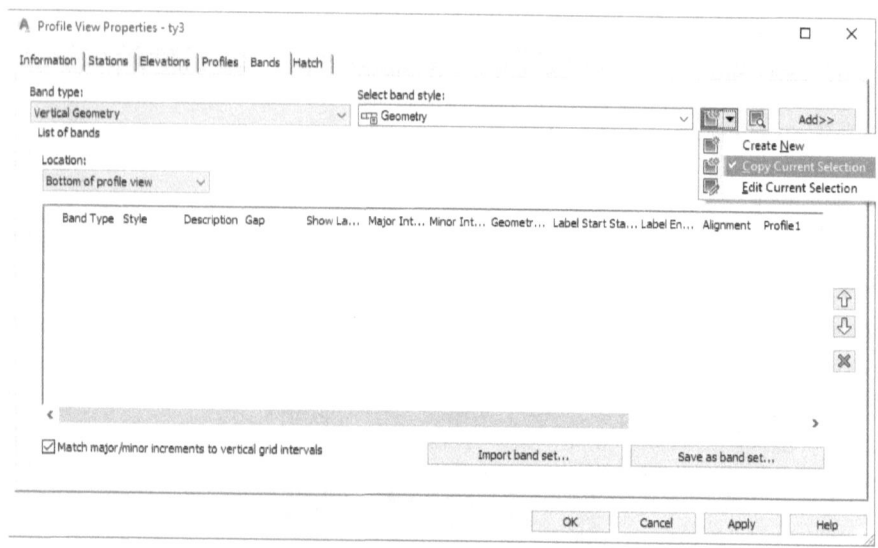

Figure E39

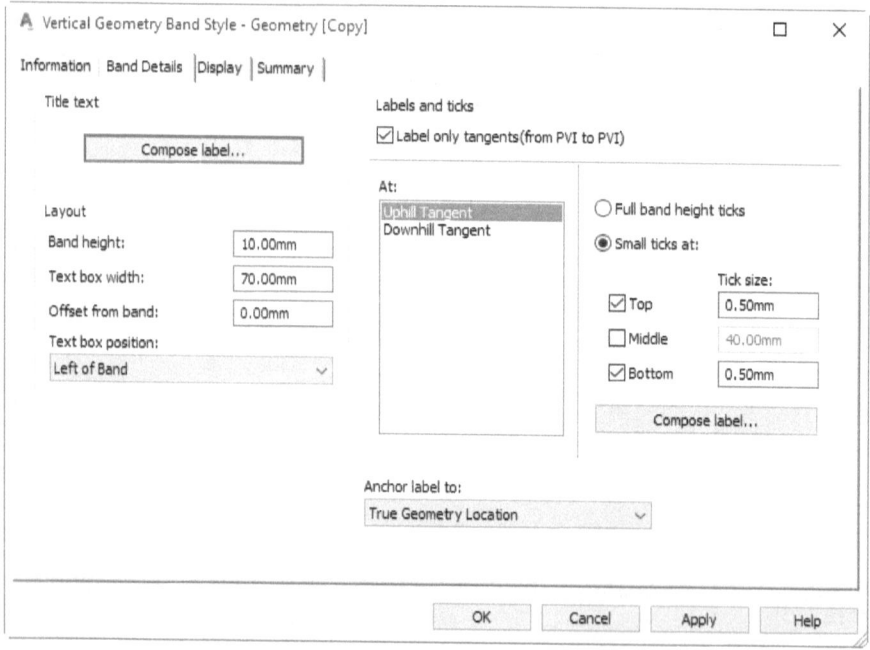

Figure E40

In the *Bands* tab of this window, set the *Band type* to *Vertical Geometry*. In the *Select band style* menu, you will have only one option to choose, namely *Geometry*. However, you can create your own band style in advance, or can click on the cascading menu to the right, select *Copy Current Selection* to create a duplicate of the current band style, and then edit it. If you choose the latter option, you will arrive at the window displayed in Figure E40.

In this window, go to the *Information* tab and type a name for your band style. Since the goal is to draw two band styles for displaying elevation and station, it is best to name the style according to its purpose (here, we named the band style *Station*). After choosing a name, go to the *Band Details* tab and check the box of *Label only tangents* option. We check this box because this profile is related to natural ground surface and does not contain uphill or downhill curves. If you want to draw the bands for a project line profile, you must uncheck this option.

Set the *Band height* to 10. To the right of this field are *Uphill Tangent* and *Downhill Tangent* options, which can be used to configure how the parameters will be displayed for positive and negative grades (uphill and downhill slopes). For example, you can adjust the settings so that station and elevation texts of positive and negative grades appear in different colors.

In the box further to the right, you can specify the way the tick of uphill and downhill tangents at intersection points will be displayed. If you select the *Full band height ticks* option, the drawn ticks will be as long as the profile band (here size 10).

Alternatively, you can select the *Small ticks at* option, and specify the length of ticks at the bottom and top of the band. In this exercise, we want to insert the elevation and station texts at the middle of the band, so you must uncheck the *Middle* option. Since the band length is 10, tick size is recommended to be 0.5. Remember that these configurations should be made for both *Uphill* and *Downhill*. Next, click on either *Uphill* or *Downhill* and press the *Compose Label* button to open the window shown in Figure E41.

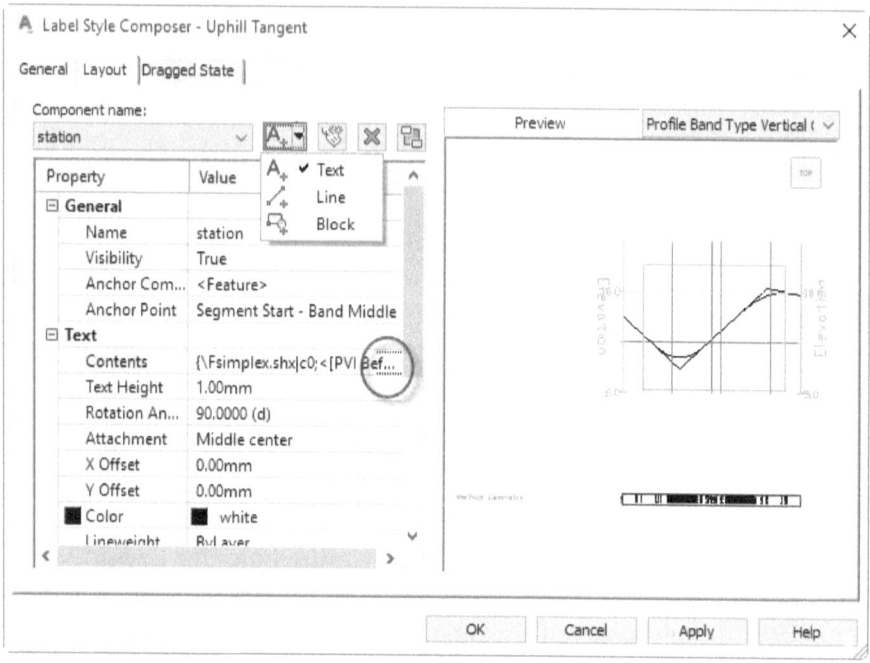

Figure E41

If there is no component in this window, press the cascading menu shown in Figure E41 and select *Text* to create a text-type component. Type a name for the component. Here, set the *Anchor Point* to *Segment Start-Band Middle* to make sure that the text appear in the middle of the band and between the two ticks defined in the previous step. Type the desired text size in the *Text Height* field, and then set the *Rotation Anchor* to 90 degrees so that the text appear vertical. Next, click on the field in front of *Contents* to open the window shown in Figure E42.

In this window, delete the contents of the right pane, then open the *Properties* menu and select *PVI Before Station*. This property will show the station (chainage) of the point where intersection starts. Press the arrow button to move the selected property to the right, adjust the font and format settings in the *Format* tab, and press OK.

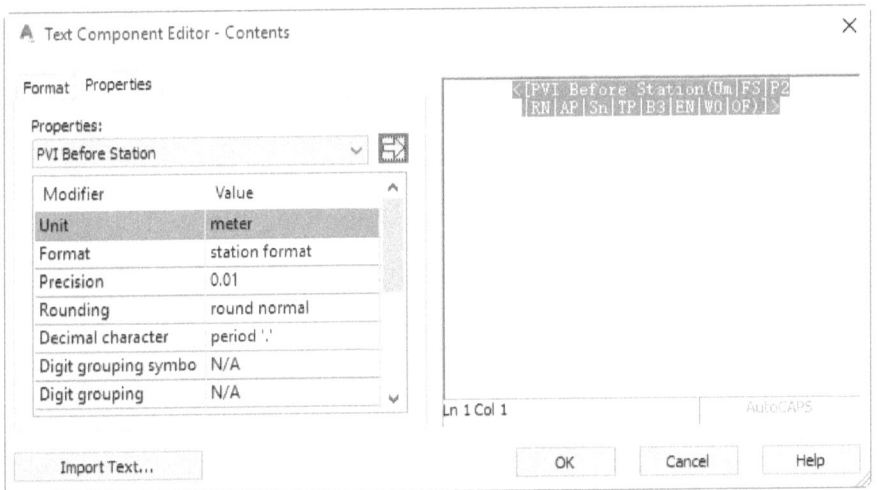

Figure E42

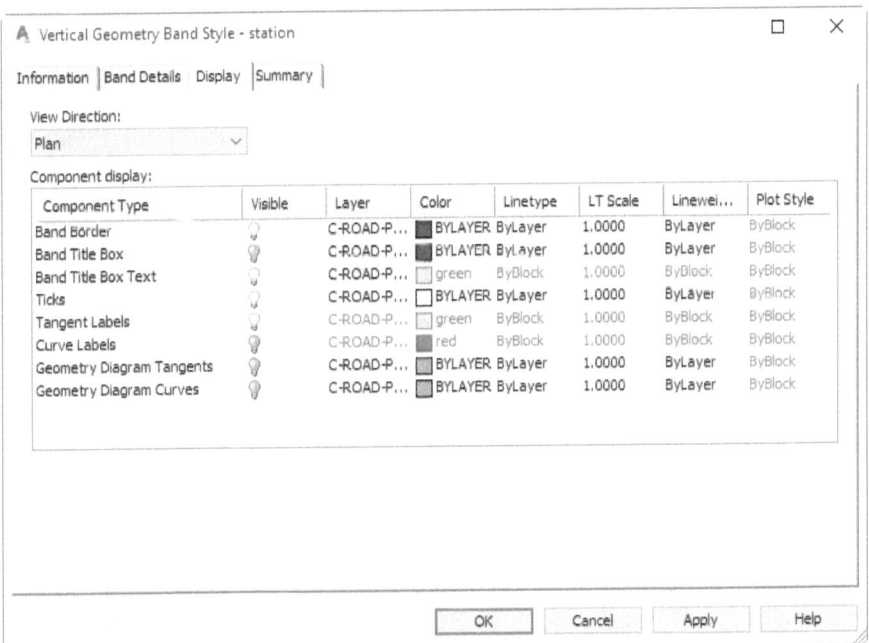

Figure E43

Also close the next window (Figure E41) to return to the window displayed in Figure E40. So far, you have instructed the software how to show stations at Uphill slopes. Now, you have to select *Downhill*, press the *Compose Label* button and repeat the above process for these slopes. When done, close the opened windows to return to the window of Figure E40. In this window, set the *Anchor label* menu to *True Geometry Location* so that station appear exactly below the intersection points. Next, go to the *Display* tab. In this tab, you can turn off the layers related to the curves and slope direction lines, as shown in Figure E43.

Once finished with these settings, press OK to return to the window shown in Figure E44.

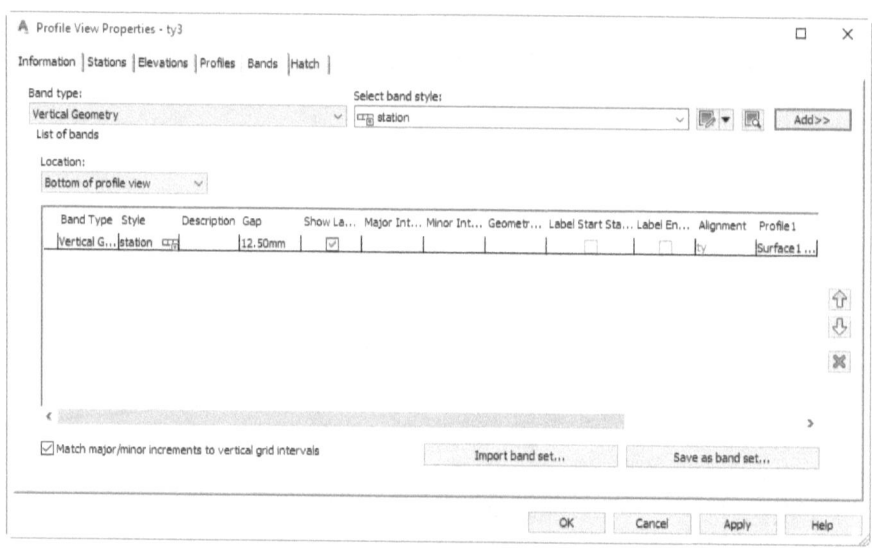

Figure E44

In this window, select the band style you just created and configured (we named it *Station*) and then press the *Add* button to apply it to the profile. After pressing OK to close the window, the *Station* band should appear under the profile.

To draw another band for elevation, again right-click on the longitudinal profile and select *Profile View Properties* to return to the window shown in Figure E44. In this window, you have to follow the above procedure to create another band style, but this time for elevations. The only difference here is that in the component assignment window you must select *PVI Before Elevation* instead of *PVI Before Station*. An easier way to do this is to create a duplicate of the band style you just created

(*Station*) and edit it to change the *PVI Before Station* to *PVI Before Elevation*.

63-How to display the slope length of profile segments above the profile or as a band under the profile?

In many projects, especially oil and gas pipeline construction projects, it is important to know the true length of the sloped segments, both over the natural ground and in the design profile. For example, for a route with a horizontal length of 100km, the true slope length along the natural ground could be 107km and the true slope length along the project line could be 104km. Given that the amount of material consumed and the work performed depends on the true length rather than the horizontal length of lines, it is imperative to include these lengths in the project drawings and estimations.

Slope length can be drawn above the profile or in a separate band under the profile. Here, we explain how to do both.

How to display slope lengths above the profile:

To do this, open the *Profiles* menu select *Add Profile View Labels* and then *Add/Edit Profile Labels* (Figure E45). The software will then ask you to select the profile. In response, click on the longitudinal profile to open the window shown in Figure E46.

In this window, set the *Type* menu to *Lines* and then click on the cascading menu next to the *Profile Tangent Label Style* and select *Create New* to start creating a new label style. This will open the window displayed in Figure E47.

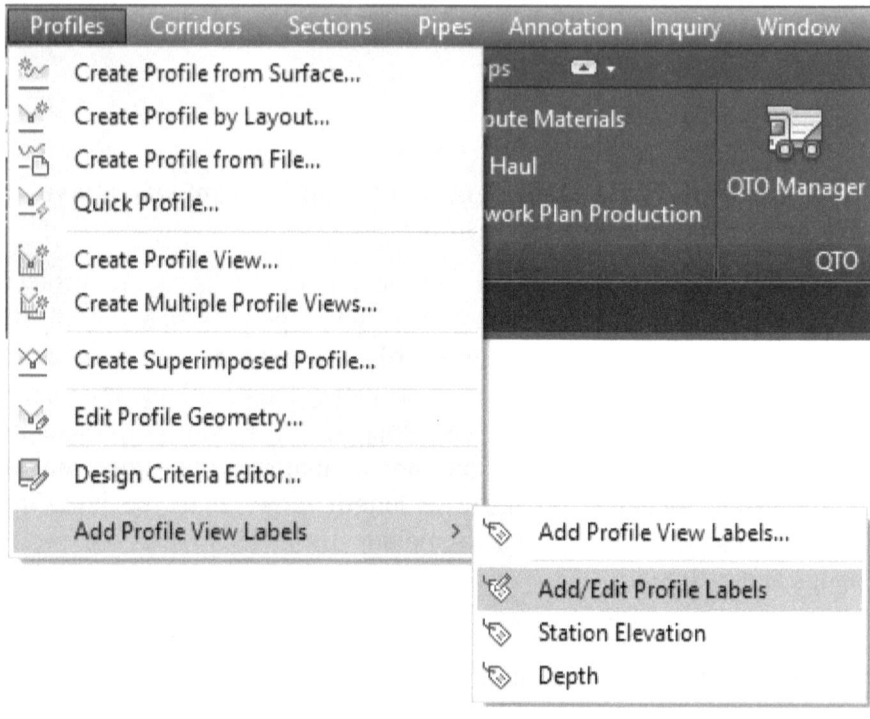

Figure E45

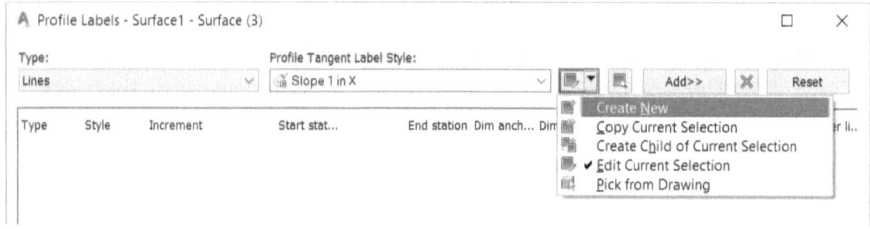

Figure E46

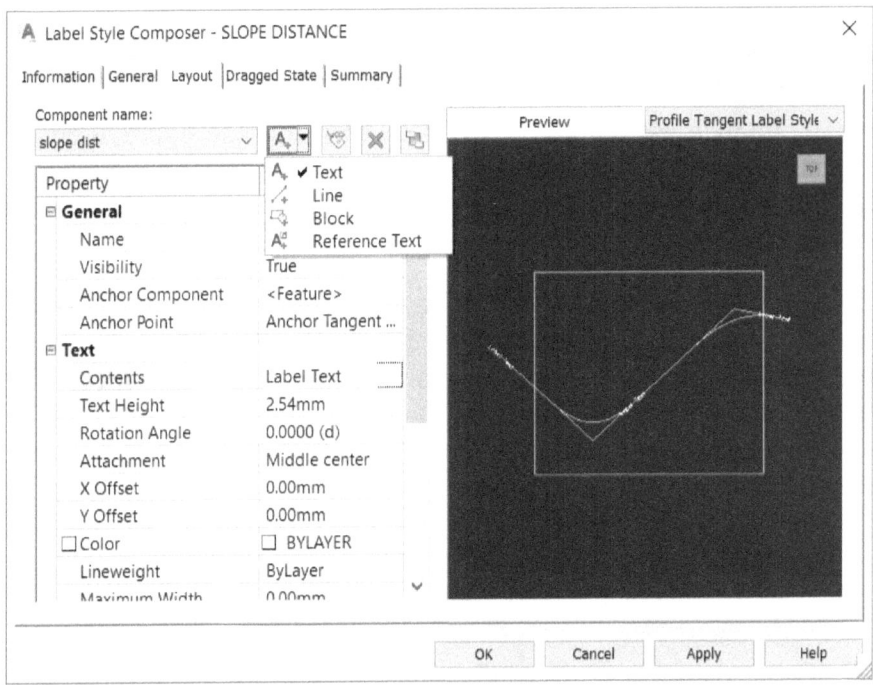

Figure E47

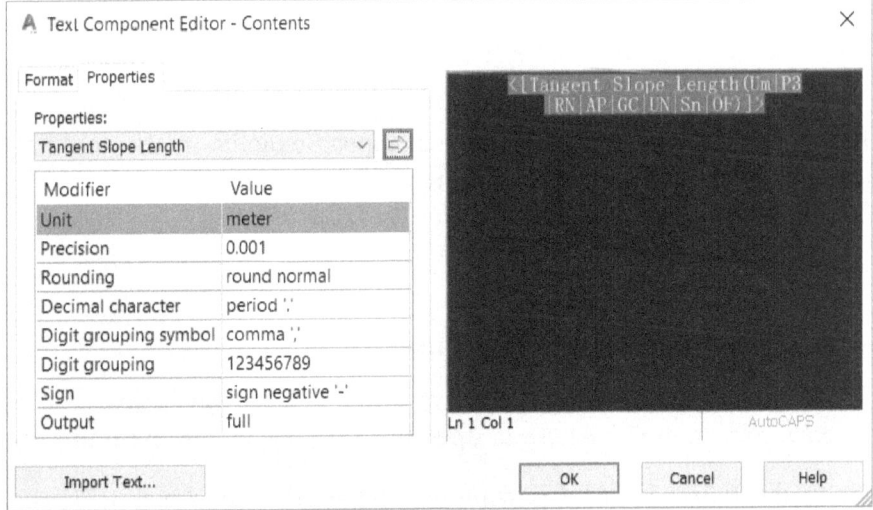

Figure E48

In this window, first go to the *Information* tab and type a name for the label to be created (here, we named it *Slope Distance*). Then, go to the *Layout* tab, use the delete button to remove the current component (Grade), and then use the cascading menu shown in the figure to create a text-type component. Type a name for the component to be created and then click on the icon in front of the *Contents* box to open the window shown in Figure E48.

In this window, remove the contents of the right pane, open the *Properties* menu and select *Tangent Slope Length*, and then press the arrow button to add it to the right pane. Adjust the appearance setting such as font and format in the *Format* tab, and then press OK to close this window. After pressing OK in the next window, you will return to the window shown in Figure E46, where you must use the *Add* button to add the created label style to the list (Figure E49). Finally, press OK to see slope length labels appear above the profile segments.

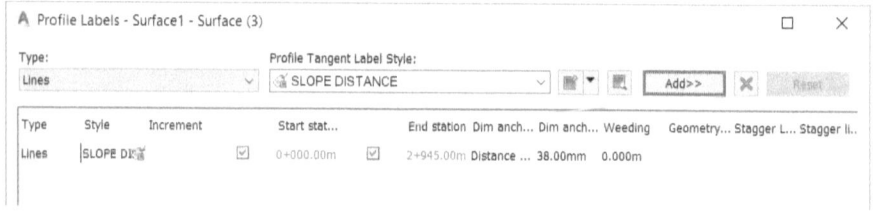

Figure E49

Remember that to make the labels appear exactly in the middle of profile segments, as shown in Figure E47, you should set the *Anchor Point* menu to *Anchor Tangent Middle*.

How to display slope lengths in a band:

To do this, select the longitudinal profile and then right-click on it and choose *Profile View Properties* to open the window shown in Figure E39.

Go to the *Bands* tab of this window, and set the *Band type* to *Vertical Geometry*. In the *Select band style* menu, you will have only one option to choose, namely *Geometry*. But you can create a band style in advance. If you have not done so, you can do it now by clicking on the cascading menu to the right. It is recommended to first select *Copy Current Selection* to create a duplicate of the currently selected band style, and then edit it. This will open the window shown in Figure E40.

Go to the *Information* tab of this window and type a name for your band style (here, we named it *Slope Distance*). Then go to the *Band Details* tab and check the box of *Label only tangents* option. This box must be ticked because this profile is related to natural ground surface and does not contain uphill or downhill curves. Uncheck this option if you want to do the same task for a project line profile.

Set the *Band height* to 10. You can use the *Uphill Tangent* and *Downhill Tangent* options in the right pane to configure how the parameters of positive and negative grades (uphill and downhill) will appear in the drawing. For example, you can use these settings to give different colors to the station and elevation texts of positive and negative grades.

In front of *Uphill Tangent* and *Downhill Tangent* options, you can specify the way the tick of uphill and downhill tangents will be drawn at intersection points. If you select the *Full band height ticks* option, the drawn ticks will be drawn with the same length as profile band (size 10). You can select the *Small ticks at* option to specify the length of ticks at the bottom and top of the band. Since elevation and station texts must appear in the middle of the band, uncheck the *Middle* option. Since the band length is 10, tick size is best to be 0.5. As in the previous exercise, these steps must be taken for both *Uphill* and *Downhill*. When done, click on *Uphill* or *Downhill* and press the *Compose Label* button to open the window shown in Figure E41.

If there is no component in this window, press the cascading menu marked in the figure and select *Text* to create a text-type component, and then type a name for the component to be created. Remember to set the *Anchor Point* menu to *Segment Mid-Band Middle* to make sure that the text appear between the start and the end of the segment. Type the desired text height in the related field, and then set the *Rotation Anchor* to 90 degrees to make the text appear vertical. Then click on the field in front of *Contents* to open the *Text Component Editor* window shown in Figure E50.

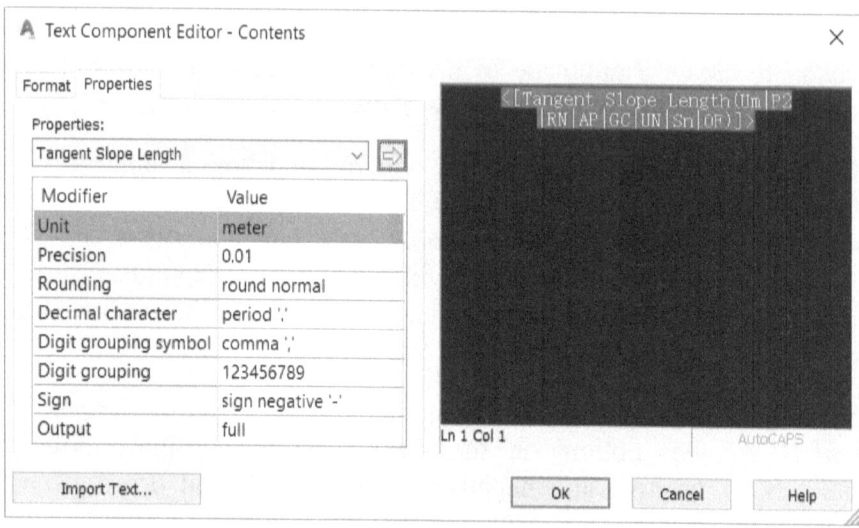

Figure E50

In this window, remove the contents of the right pane, open the *Properties* menu and select *Tangent Slope Length*, and press the arrow button to move the selected property to the right. After adjusting the font and format settings in the *Format* tab, press OK to close the window. Press OK in the next window (Figure E41) to return to the window shown in Figure E40. So far, you have instructed the software how to show the slope length at Uphill slopes. Now, you have to select *Downhill*, press the *Compose Label* button and repeat the above process for downhill slopes as well. Once finished, press OK to close the opened windows and return to this very same window (Figure E40). Now, set the *Anchor label* menu to *True Geometry Location* so that slope length labels appear exactly where you want. Then, go to the *Display* tab and turn off the layers related to the curves and slope direction lines, as shown in Figure E43.

Once finished with these settings, press OK to return to the window shown in Figure E44.

In this window, select the band style you just created (*Slope Distance*) and then press the *Add* button to apply it to the profile. Press OK to see the new band appear under the profile.

64-How to create a band for displaying the grade between the segments of a longitudinal profile?

This can be done in the same way as explained in the second part of the previous exercise, except that in the window of Figure E50, you must select *Tangent Grade* from the *properties* menu.

65-How to create a band for displaying the true chainage at the intersection points of a longitudinal profile?

People who have worked with the software SDR Map may know that this software allows the user to insert the true chainage (as opposed to 2-dimnetional/horizontal chainage) of points into profile drawings. But to the best our knowledge, this feature is not provided in Civil3D (V. 2018). Considering the importance of this issue for many designers, here, we explain how to create a band for displaying the true chainage values using Civil3D, AutoCAD and Excel.

The first step is to create an empty band containing only the intersection points of the longitudinal profile.

To do this, select the longitudinal profile and then right-click on it and select *Profile View Properties* to open the window shown in Figure E39.

Go to the *Bands* tab and set the *Band type* to *Vertical Geometry*. In the *Select band style* menu, you will have only one option to choose (*Geometry*), but you can create your own band style in advance. If you have not done this earlier, click on the cascading menu to the right, select *Copy Current Selection* to create a duplicate band style, and then edit it. This will open the window shown in Figure E40.

In this window, go to the *Information* tab and type a name for the band style to be created (here, we named it *Slope KM*). Then, go to the *Band Details* tab and check the box of *Label only tangents* option. As explained in previous exercises, we check this box because this profile is related to natural ground surface and does not contain uphill or downhill curves. You can use this option if you want to do this task for a project line profile.

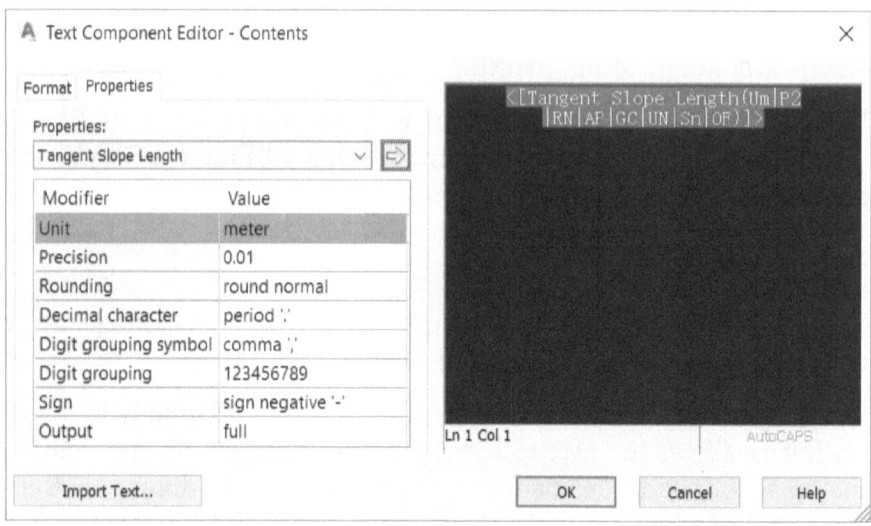

Figure E51

The settings provided to the right of *Uphill Tangent* and *Downhill Tangent* options allow you to specify the way the tick will be drawn at intersection points. If you select the *Full band height ticks* option, the ticks will be drawn with the same length as profile band (size 10). You can select the *Small ticks at* option to specify the length of ticks at the bottom and top of the band. Since we want the true chainage to appear in the middle of the band, check the *Top* and *Bottom* options and uncheck the *Middle* option. Given that the band length is 10, set the tick size to 0.5. Remember that these steps must be taken for both *Uphill* and *Downhill* options. The next step is to create the empty band. For this purpose, click on *Uphill* and press the *Compose Label* button to open the window shown in Figure E51.

As shown in Figure E50, delete all the components in this window and press OK. Do the same for Downhill. After returning to the window shown in Figure E44, select the created empty band style and then press the *Add* button to apply it to the profile. Press OK to see a band containing only the intersection points appear under the profile (Figure E52). The next step is to insert the true chainage values into this empty band.

Figure E52

To obtain these values, you need to export an Excel file containing the intersection points and their respective elevations, grades and other parameters needed to calculate the true chainage.

To calculate the true chainage of a point from an adjacent reference point, you will need the elevation difference of the two points and the horizontal distance between them. For example, consider the profile segment displayed in Figure E53.

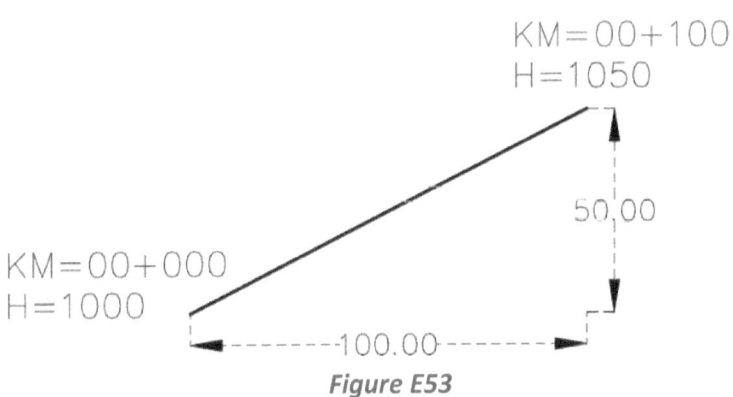

KM=00+100
H=1050

50,00

KM=00+000
H=1000

100.00

Figure E53

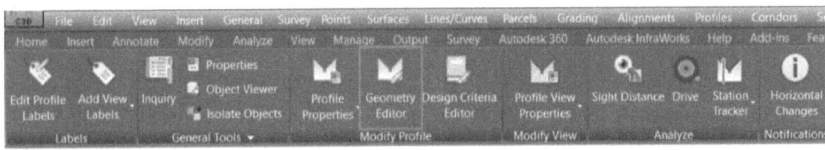

Figure E54

In this profile segment, the left point is located at the station 00+000 at an elevation of 1000 and the right point is located at the station 00+100 at an elevation of 1050. Therefore, the two points have an elevation difference of 50 and a horizontal distance of 100 meters.

At the first point, the horizontal chainage and the true chainage are equal (00+000), but the same cannot be said for the second point.

Thus, we need to calculate the true chainage of the second point, which equals the chainage of the first point plus the slope length. According to the Pythagorean Theorem, the slope length can be obtained from the following equation.

SD=((100^2)+(50^2))^0.5 = 111.80

Thus, the slope distance between the two points is 111.80 meters, which when summed with the chainage of the first point (00+000), give the true chainage of the second point as 00+111.80. This true chainage can be used to obtain the true chainage of a third point located to the right of the second point. The true chainage of a series of consecutive points can be calculated in the same way.

Hence, to calculate the true chainage of a series of points, we need the true chainage of a reference point, the horizontal stationing of the points and their elevations. Thus, we have to create an Excel file containing a list of these parameters.

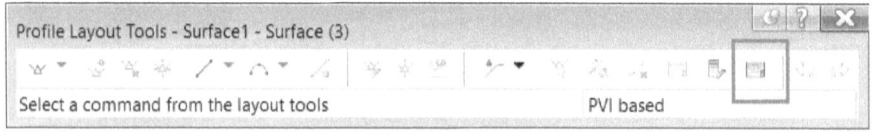

Figure E55

To export this file from a Civil3D drawing, select the profile line and then press the *Geometry Editor* button on the ribbon (Figure E54) to open the window displayed in Figure E55.

In this window, click on the button marked in the figure to open the profile geometry table shown in Figure E56.

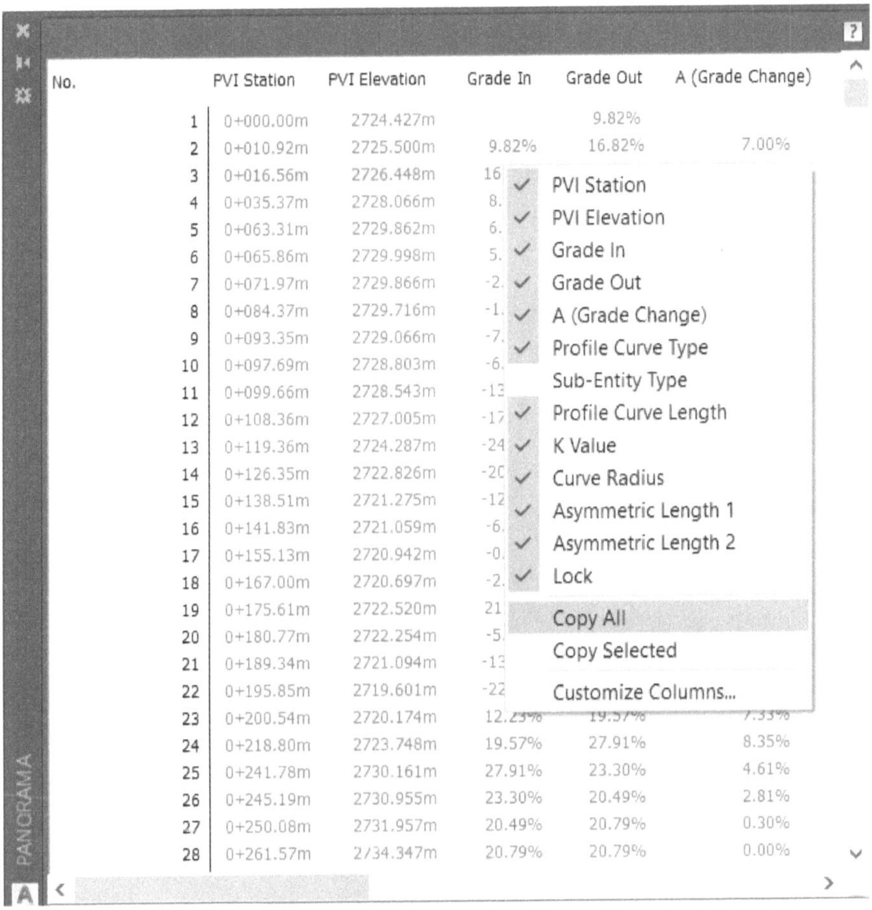

No.	PVI Station	PVI Elevation	Grade In	Grade Out	A (Grade Change)
1	0+000.00m	2724.427m		9.82%	
2	0+010.92m	2725.500m	9.82%	16.82%	7.00%
3	0+016.56m	2726.448m	16.		
4	0+035.37m	2728.066m	8.		
5	0+063.31m	2729.862m	6.		
6	0+065.86m	2729.998m	5.		
7	0+071.97m	2729.866m	-2.		
8	0+084.37m	2729.716m	-1.		
9	0+093.35m	2729.066m	-7.		
10	0+097.69m	2728.803m	-6.		
11	0+099.66m	2728.543m	-13		
12	0+108.36m	2727.005m	-17		
13	0+119.36m	2724.287m	-24		
14	0+126.35m	2722.826m	-20		
15	0+138.51m	2721.275m	-12		
16	0+141.83m	2721.059m	-6.		
17	0+155.13m	2720.942m	-0.		
18	0+167.00m	2720.697m	-2.		
19	0+175.61m	2722.520m	21		
20	0+180.77m	2722.254m	-5.		
21	0+189.34m	2721.094m	-13		
22	0+195.85m	2719.601m	-22		
23	0+200.54m	2720.174m	12.23%	19.57%	7.33%
24	0+218.80m	2723.748m	19.57%	27.91%	8.35%
25	0+241.78m	2730.161m	27.91%	23.30%	4.61%
26	0+245.19m	2730.955m	23.30%	20.49%	2.81%
27	0+250.08m	2731.957m	20.49%	20.79%	0.30%
28	0+261.57m	2734.347m	20.79%	20.79%	0.00%

Context menu (overlaying the table):
- ✓ PVI Station
- ✓ PVI Elevation
- ✓ Grade In
- ✓ Grade Out
- ✓ A (Grade Change)
- ✓ Profile Curve Type
- Sub-Entity Type
- ✓ Profile Curve Length
- ✓ K Value
- ✓ Curve Radius
- ✓ Asymmetric Length 1
- ✓ Asymmetric Length 2
- ✓ Lock
- Copy All
- Copy Selected
- Customize Columns...

Figure E56

As you can see, this table reports the horizontal stationing of the intersection points, their elevations, and the slope of the lines coming in and going out of each intersection point (Grade In and Grade Out).

To export this data to Excel, just right-click on one of the cells in the table and select *Copy All*, then open a blank Excel document and paste it there. Now that you have a complete report of the longitudinal profile in Excel format, you just have to incorporate the slope distance calculations into this file.

As mentioned earlier, the parameters needed for these calculations are the elevation difference and the horizontal distance, so you can remove all columns except stationing and elevation to reach a simpler list as shown in Figure E57.

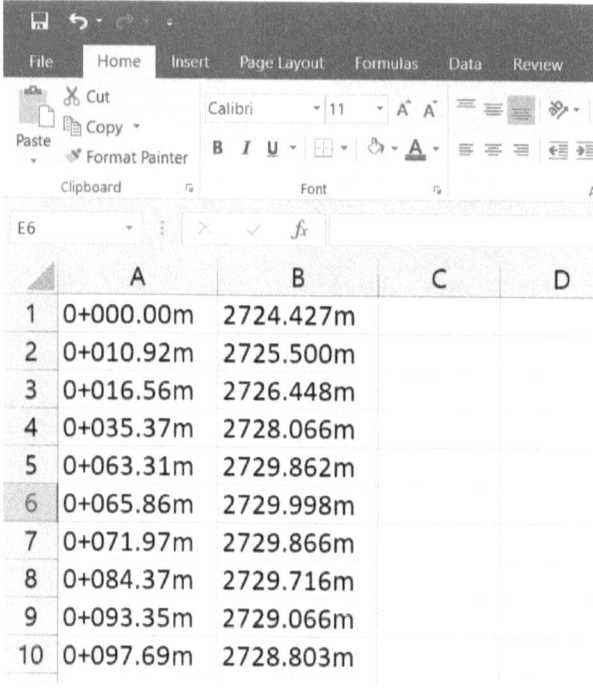

Figure E57

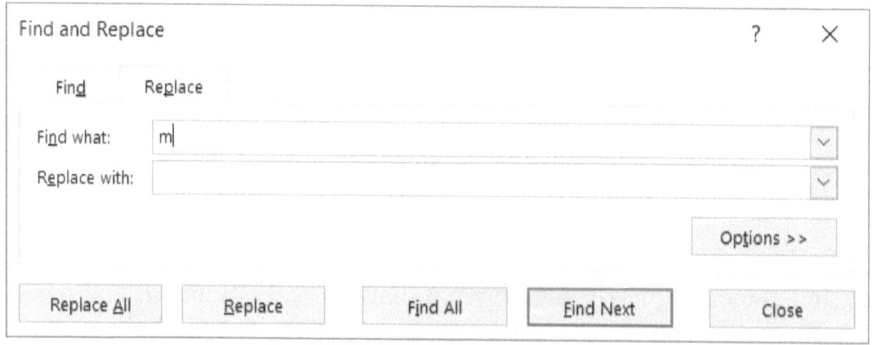

Figure E58

These two columns contain not only numbers but also characters such as *m* and +, which must be removed before any function can be defined for their contents. For this purpose, select the column B and press Ctrl + H to open the window shown in Figure E58.

In this window, open the *Replace* tab, type the character you want to remove in the *Find what* field, leave the *Replace with* field empty (since you want the character removed), and press the *Replace All* button.

Do the same for column A and character +. Now, the file is ready for defining functions.

	A	B	C	D
1	0	2724.427		
2	10.92	2725.5	=A2-A1	=B2-B1
3	16.56	2726.448		
4	35.37	2728.066		
5	63.31	2729.862		
6	65.86	2729.998		
7	71.97	2729.866		
8	84.37	2729.716		
9	93.35	2729.066		
10	97.69	2728.803		
11	99.66	2728.543		
12	108.36	2727.005		
13	119.36	2724.287		
14	126.35	2722.826		
15	128.51	2721.275		

Figure E59

Here, you need to create two columns, one for calculating the elevation difference and the other for calculating the horizontal distance of stations. As shown in Figure E59, go to the second row of column C (cell C2), type =A2-A, and press enter, then go to the second row of column D (cell D2) and type =B2-B1 and press enter. When done, select the cell C2, move the mouse pointer to the bottom right corner of this cell until the cursor changes to a plus sign, and then click and drag down until reaching the last row of data. In doing so, you will instruct the software to reproduce the function for all the rows below the one originally selected. Repeat this process for the cell D2. The next step is to prepare column E for calculating the slope length based on the Pythagorean Theorem $(C^2=A^2+B^2)$.

As shown in Figure E60, go to cell E2, type =((C2^2)+(D2^2))^.5 and press enter.

	A	B	C	D	E
1	0	2724.427			
2	10.92	2725.5	10.92	1.073	=((C2^2)+(D2^2))^0.5
3	16.56	2726.448	5.64	0.948	
4	35.37	2728.066	18.81	1.618	
5	63.31	2729.862	27.94	1.796	
6	65.86	2729.998	2.55	0.136	
7	71.97	2729.866	6.11	-0.132	
8	84.37	2729.716	12.4	-0.15	
9	93.35	2729.066	8.98	-0.65	
10	97.69	2728.803	4.34	-0.263	
11	99.66	2728.543	1.97	-0.26	
12	108.36	2727.005	8.7	-1.538	
13	119.36	2724.287	11	-2.718	
14	126.35	2722.826	6.99	-1.461	
15	138.51	2721.275	12.16	-1.551	
16	141.83	2721.059	3.32	-0.216	
17	155.13	2720.942	13.3	-0.117	
18	167	2720.697	11.87	-0.245	

Figure E60

By doing so, you instruct the software to calculate the squares of the elevation difference and the horizontal distance, sum up the answers, and raise this sum to the power of 0.5 to obtain its square root.

Repeat the above-explained click and drag process for the cell E2 to let the software calculate the slope length for all listed points. The next step is to calculate the true chainage. As shown in Figure E61, type 0 in the cell F1 (this point acts as the reference), then go to cell F2, and type =F1+E2 and press enter. This means that to calculate the true chainage of each point, the true chainage of its previous point should be summed with the slope distance of the two points. Click and drag this cell as explained above to make the software calculate the true chainage of all points according to the same formula (Figure E62).

	A	B	C	D	E	F	G
1	0	2724.427					0
2	10.92	2725.5	10.92	1.073	10.97	=F1+E2	
3	16.56	2726.448	5.64	0.948	5.72		
4	35.37	2728.066	18.81	1.618	18.88		
5	63.31	2729.862	27.94	1.796	28.00		
6	65.86	2729.998	2.55	0.136	2.55		
7	71.97	2729.866	6.11	-0.132	6.11		
8	84.37	2729.716	12.4	-0.15	12.40		
9	93.35	2729.066	8.98	-0.65	9.00		
10	97.69	2728.803	4.34	-0.263	4.35		

Figure E61

	A	B	C	D	E	F	
1	0	2724.427				0	
2	10.92	2725.5	10.92	1.073	10.97	10.97	
3	16.56	2726.448	5.64	0.948	5.72	16.69	
4	35.37	2728.066	18.81	1.618	18.88	35.57	
5	63.31	2729.862	27.94	1.796	28.00	63.57	
6	65.86	2729.998	2.55	0.136	2.55	66.12	
7	71.97	2729.866	6.11	-0.132	6.11	72.23	
8	84.37	2729.716	12.4	-0.15	12.40	84.63	
9	93.35	2729.066	8.98	-0.65	9.00	93.64	
10	97.69	2728.803	4.34	-0.263	4.35	97.99	
11	99.66	2728.543	1.97	0.26	1.99	99.97	
12	108.36	2727.005	8.7	-1.538	8.83	108.81	
13	119.36	2724.287	11	-2.718	11.33	120.14	
14	126.35	2722.826	6.99	-1.461	7.14	127.28	
15	138.51	2721.275	12.16	-1.551	12.26	139.54	
16	141.83	2721.059	3.32	-0.216	3.33	142.87	
17	155.13	2720.942	13.3	-0.117	13.30	156.17	
18	167	2720.697	11.87	-0.245	11.87	168.04	

Figure E62

As you can see, the results will be the automatic calculation of the true chainage for all points. For example, point No.15 located at the station (horizontal chainage) of 00+138.51 has a true chainage of 00+139.54.

Now that you have obtained the true chainage values, you have to insert these value into the empty band created earlier. But before doing this, these values must be transformed into a suitable station format. For this purpose, right-click on the true chainage column and select *Format Cells* option to open the window shown in Figure E63.

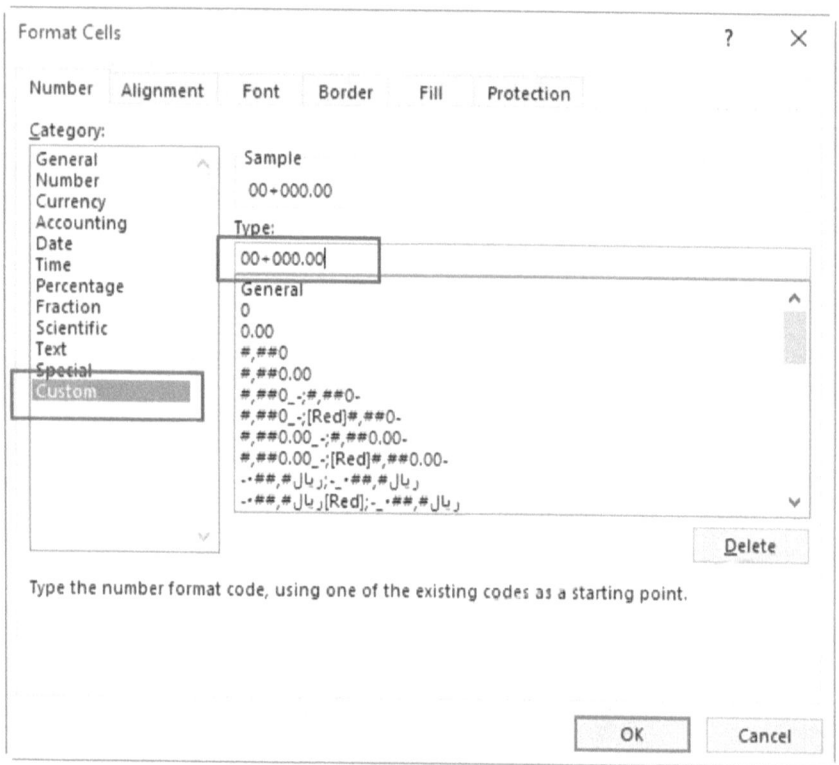

Figure E63

Go to the *Number* tab of the opened window, select *Custom* in the *Category* box, type 00+000.00 in the *Type* field and press OK to close the window. This makes sure that the chainage values will be displayed in the standard station format. But this is not enough, because although the points are converted into correct format, when imported into AutoCAD, they will not be displayed in the desired manner. To resolve this issue, you must convert this excel file to a Text (Tab Delimited) file and then re-open it with Excel.

Now, it is time to insert these true chainage values into the created band.

However, since these value should be inserted as text, you need the coordinates of the insertion points.

The band is positioned horizontally, so all the points on this band have the same Y coordinate. Thus, you can use the ID command to obtain the coordinates of the first point, and use its Y coordinate for every insertion point. Also, the X coordinates of the insertion points are directly related to their horizontal distance. Hence, you can use the ID command to obtain the X coordinate of the first insertion point, and then sum it with the available horizontal distances to calculate the X coordinate of other points.

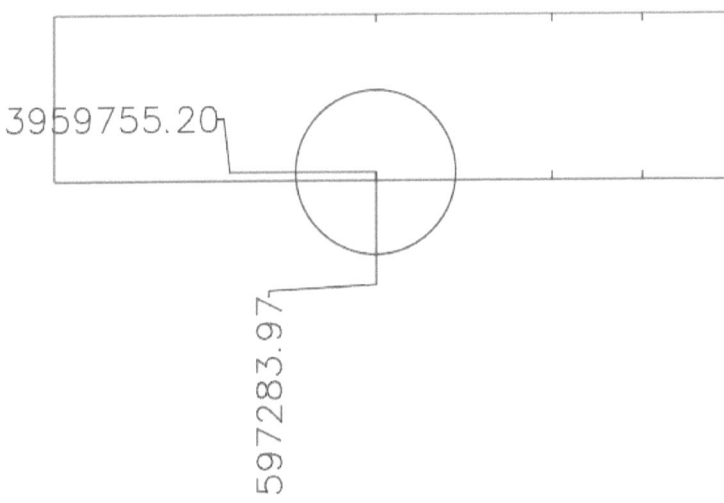

Figure E64

As shown in Figure E64, using the ID command on the first insertion point gives the X coordinate as 597283.97 and the Y coordinate as 3959755.20. Reopen to the Excel file and copy these values into two columns as shown in Figure E65 (we used columns G and H for X and Y coordinates of the insertion points). As stated earlier, all insertion points have the same Y coordinate, so just copy this value at all rows of the column H. The X value, however, increases with the horizontal distance, so go to the cell G3, type =, select the cell G2 and press the F4 key to fix it (it should turn into $G$2), then type +, select the cell A3 and press enter. The end result should be the formula shown in Figure E65.

	A	B	C	D	E	F	G	H
							X	Y
1								
2	0	2724.427				00+000.00	597283.97	3959755.21
3	10.92	2725.5	10.92	1.073	10.97	00+010.97		
4	16.56	2726.448	5.64	0.948	5.72	00+016.69		
5	35.37	2728.066	18.81	1.618	18.88	00+035.57		
6	63.31	2729.862	27.94	1.796	28	00+063.57		
7	65.86	2729.998	2.55	0.136	2.55	00+066.12		
8	71.97	2729.866	6.11	-0.132	6.11	00+072.23		
9	84.37	2729.716	12.4	-0.15	12.4	00+084.63		
10	93 35	2729 066	8 98	-0 65	9	00+093 64		

Figure E65

The purpose of pressing the F4 key (using $G$2 instead of G2) is that we want all horizontal distances to be added to the coordinate of this cell, so we must use $G$2 to make sure that the cell G2 remains a constant component of the formula as we drag down the cell G3 to make the software reproduce the formula for the next rows of that column (Figure E66).

	A	B	C	D	E	F	G	H
							X	Y
1								
2	0	2724.427				00+000.00	597283.97	3959755.21
3	10.92	2725.5	10.92	1.073	10.97	00+010.97	=$G$2+A3	
4	16.56	2726.448	5.64	0.948	5.72	00+016.69		
5	35.37	2728.066	18.81	1.618	18.88	00+035.57		
6	62 21	2729 962	27 94	1 796	29	00+062 57		

Figure E66

At the end of this step, the file should contain the X and Y coordinates of all insertion points as shown in Figure E67.

	A	B	C	D	E	F	G	H	I
1							X	Y	
2	0	2724.427				00+000.00	597283.97	3959755.21	
3	10.92	2725.5	10.92	1.073	10.97	00+010.97	597294.89	3959755.21	
4	16.56	2726.448	5.64	0.948	5.72	00+016.69	597300.53	3959755.21	
5	35.37	2728.066	18.81	1.618	18.88	00+035.57	597319.34	3959755.21	
6	63.31	2729.862	27.94	1.796	28	00+063.57	597347.28	3959755.21	
7	65.86	2729.998	2.55	0.136	2.55	00+066.12	597349.83	3959755.21	
8	71.97	2729.866	6.11	-0.132	6.11	00+072.23	597355.94	3959755.21	
9	84.37	2729.716	12.4	-0.15	12.4	00+084.63	597368.34	3959755.21	
10	93.35	2729.066	8.98	-0.65	9	00+093.64	597377.32	3959755.21	
11	97.69	2728.803	4.34	-0.263	4.35	00+097.99	597381.66	3959755.21	
12	99.66	2728.543	1.97	-0.26	1.99	00+099.97	597383.63	3959755.21	
13	108.36	2727.005	8.7	-1.538	8.83	00+108.81	597392.33	3959755.21	
14	119.36	2724.287	11	-2.718	11.33	00+120.14	597403.33	3959755.21	
15	126.35	2722.826	6.99	-1.461	7.14	00+127.28	597410.32	3959755.21	
16	138.51	2721.275	12.16	-1.551	12.26	00+139.54	597422.48	3959755.21	
17	141.83	2721.059	3.32	-0.216	3.33	00+142.87	597425.8	3959755.21	

Figure E67

To display the true chainage values at the corresponding coordinates, it is easier to use a *Concatenate* function containing a *-text* command. For this purpose, click on the cell I2, click on the f_x button above and select the *Concatenate* function from the list.

I2				f_x	=CONCATENATE("-text"," ",G2,"," ,H2," ",5," ",90," ",F2)			
	A	B	C	D	E	F	G	H	I
1							X	Y	
2	0	2724.427				00+000.00	597283.97	3959755.21	-text 597283.97,3959755.205 5 90 00+000.00
3	10.92	2725.5	10.92	1.073	10.97	00+010.97	597294.89	3959755.21	
4	16.56	2726.448	5.64	0.948	5.72	00+016.69	597300.53	3959755.21	
5	35.37	2728.066	18.81	1.618	18.88	00+035.57	597319.34	3959755.21	
6	63.31	2729.862	27.94	1.796	28	00+063.57	597347.28	3959755.21	
7	65.86	2729.998	2.55	0.136	2.55	00+066.12	597349.83	3959755.21	
8	71.97	2729.866	6.11	-0.132	6.11	00+072.23	597355.94	3959755.21	

Figure E68

Setup the *Concatenate* function as shown in Figure E68, then copy all contents of this column and paste it into the AutoCAD command line (for the file containing the longitudinal profile). After running this command, the true chainage values will appear on the band of the longitudinal profile.

66-How to display true chainage at every intersection point?

In the previous exercise, we learned how to calculate the true chainage values based on the data extracted from a longitudinal profile and how to insert them into a band in the profile drawing. In this exercise, the goal is make the true chainage values appear not in a separate band under the profile but on the profile itself.

As you can guess, the only difference between this exercise and the previous one is the point of insertion. In fact, the X-coordinate of the insertion points must also be obtained as explained in the previous exercise, and the only difference is in the Y-coordinate of these points. The only problem is that because of the vertical exaggeration that is typical to longitudinal profiles, the elevation line is often several times (usually10 times) higher or lower than where it would be without this exaggeration. This is somewhat problematic because we want the true chainage values appear right above the elevation line.

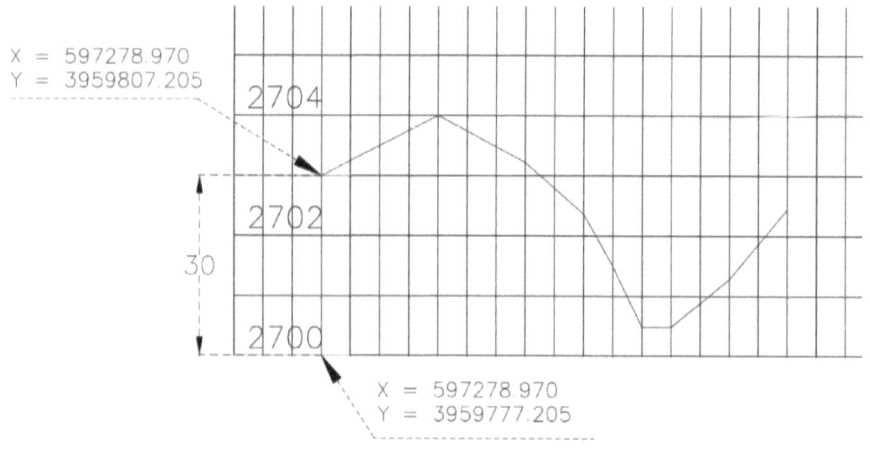

Figure E69

When the exaggeration ratio is 1, the Y-coordinates of the insertion points can be obtained by subtracting the elevation of each point from the elevation of the reference point and then summing the answers with the Y-coordinate of the reference point.

Consider the following example.

Figure E69 shows a segment of a longitudinal profile, where the reference elevation is 2700 and the Y-coordinate of the reference line is 3959777.205. The first point of the profile has an elevation of 2703. Thus, to obtain the Y-coordinate of this point, the elevation of this point (2703) should be subtracted from the reference elevation (2700). Given that the profile is drawn with a vertical exaggeration ratio of 10, the answer of the above subtraction (3) should be multiplied by 10, and the answer (30) should be added to the Y-coordinate of the reference point to get the coordinate of the insertion point (3958807.205). The Y coordinates of all insertion points can be calculated in the same way.

	A	B	C	D	E
1				Profile Elevation Datum	2700
2				X datum point	597278.97
3				Y datum point	3959777.205
4					
5	Horizantal KM	Elevation	Slope KM	X insertion point	Y insertion point
6	00+000.00	2724.427	00+000.00		
7	00+010.92	2725.5	00+010.97		
8	00+016.56	2726.448	00+016.69		
9	00+035.37	2728.066	00+035.57		
10	00+063.31	2729.862	00+063.57		

Figure E70

As in previous exercise, we first calculate the true chainage values and then obtain the insertion points. But first, rearrange the Excel file created for the true chainage to the form shown in Figure E70.

To calculate the X coordinate of the insertion points, you have to instruct the software to sum the X coordinate of the reference point, i.e. the cell E2, with the horizontal station value (see Figure E71). Remember to use the F4 key to change E2 to $E$2 when writing the formula.

	A	B	C	D	E
1				Profile Elevation Datum	2700
2				X datum point	597278.97
3				Y datum point	3959777.205
4					
5	Horizantal KM	Elevation	Slope KM	X insertion point	Y insertion point
6	00+000.00	2724.427	00+000.00	=$E$2+A6	
7	00+010.92	2725.5	00+010.97		
8	00+016.56	2726.448	00+016.69		
9	00+035.37	2728.066	00+035.57		
10	00+063.31	2729.862	00+063.57		

Figure E71

Now, to calculate the Y coordinate of the insertion points, you must write a formula in the cell E6 for summing the Y-coordinate of the reference points ($E$3) with the difference between the reference elevation ($E$1) and the elevation of the point (B6) multiplied by 10. This formula is shown in Figure E72.

	A	B	C	D	E
1				Profile Elevation Datum	2700
2				X datum point	597278.97
3				Y datum point	3959777.205
4					
5	Horizantal KM	Elevation	Slope KM	X insertion point	Y insertion point
6	00+000.00	2724.427	00+000.00	597278.97	=$E$3+10($E$1-B6)
7	00+010.92	2725.5	00+010.97		
8	00+016.56	2726.448	00+016.69		
9	00+035.37	2728.066	00+035.57		
10	00+063.31	2729.862	00+063.57		

Figure E72

When done creating the formulas, click and drag down the cells to make the software reproduce the formulas for all rows. The next step is to use the *Concatenate* function to create the formula shown in Figure E73 as explained in the previous exercise. Drag the created cell over the entire column F to make the software autofill the rows accordingly.

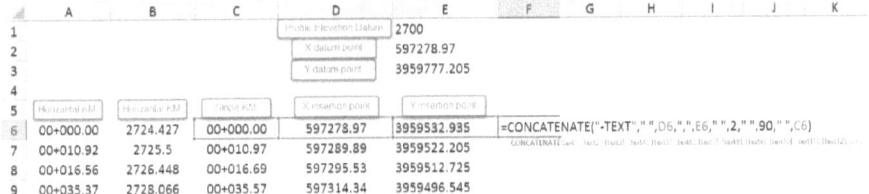

Figure E73

Finally, copy this column and paste it into the command line of the profile file. The end result of this process should be the true chainage values appearing over the profile, as shown in Figure E74.

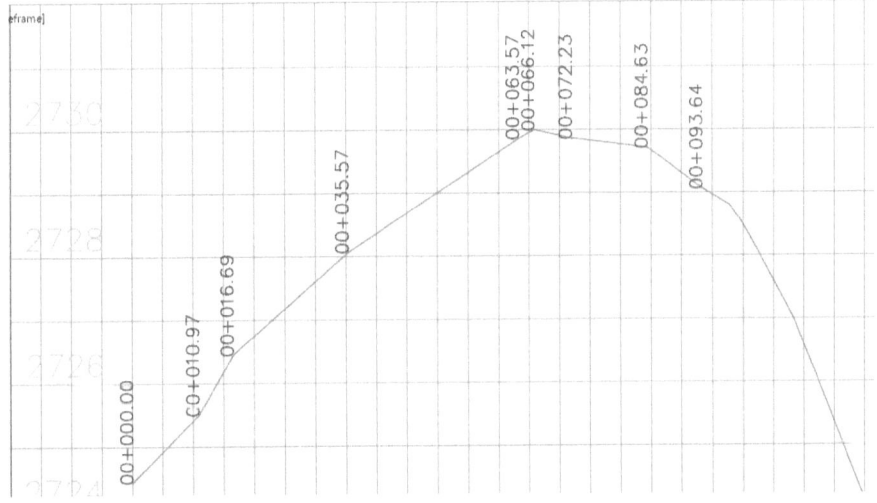

Figure E74

67-How to draw a longitudinal profile only with the route's axis points?

One of the common problems of engineers working in the road and pipeline construction projects is how to plot a longitudinal profile based on the route's axis points. As you know, longitudinal profiles must be derived from a surface, and since one cannot create a surface from a survey of axis points, such survey information cannot be used to produce a profile in a standard way.

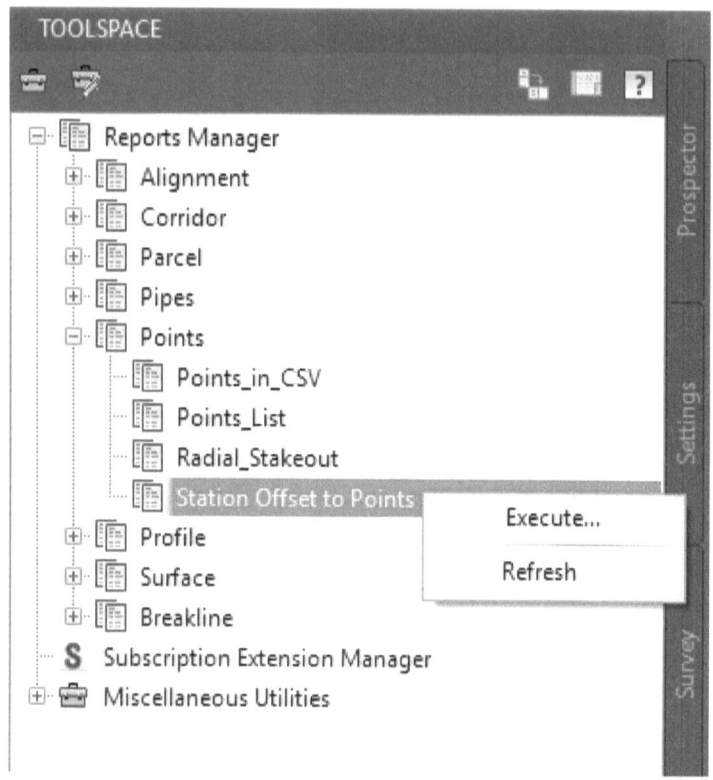

Figure E75

To overcome this challenge, we need a file of surveyed points with the station-elevation format, that is, with the stations of the points given in one column and the corresponding elevations given in the other column. The process will be easier if the survey data are already available in such a format, but if they are just coordinates, they should first be converted into a format that is processable by the software, i.e. the station-elevation format

In this exercise, we first explain how to convert a set of coordinates to the station-elevation format, and then describe how the results can be used to drawing a profile.

For this exercise, open the file <u>Profile-on-Point.dwg</u> from the folder <u>Project File</u>. This file contains a route (alignment) and the points surveyed along its axis. One of the advantages of converting the coordinates to the station-elevation format is that all offsets can be considered zero even if the points are not exactly on the axis. Note that it is impossible to reach a zero offset even if we set out the points and survey them again.

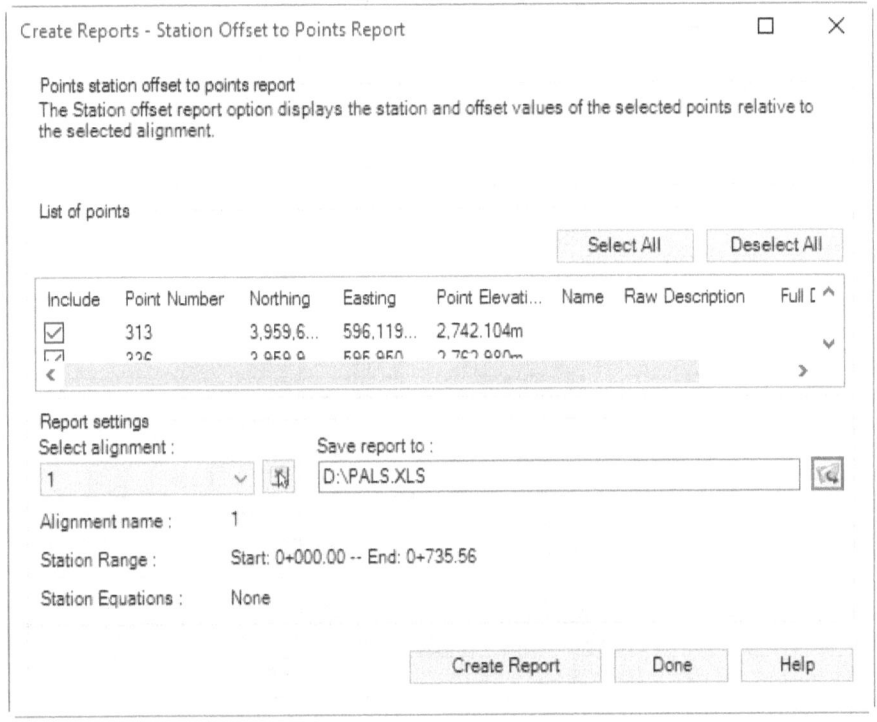

Figure E76

The first step is to convert this file into the station-elevation format.

For this purpose, open the *Toolbox* in *TOOLSPACE* window and expand the *Reports Manager branch* and then *Points* sub-branch. Right-click on *Station Offset to Points* and select *Execute* (Figure E75). This will open the window shown in Figure E76.

In this window, press the *Select All* button to select all points, then open the *Select Alignment menu*, and select the alignment according to which the points must be extracted. Specify the path and format of the output file in the *Save report to* field and then press the *Create Report* button to generate the file.

	A	B	C	D	E
13					
14	Point	Station	Offset	Elevation	Description
15	313	0+067.43	-0.519m	2,742.104m	
16	336	0+374.81	0.707m	2,762.980m	
17	312	0+454.45	-0.043m	2,773.310m	
18	314	0+475.32	-0.231m	2,770.194m	
19	331	0+395.13	0.598m	2,768.602m	
20	323	0+422.06	0.300m	2,774.133m	
21	471	0+540.89	0.043m	2,770.472m	
22	343	0+342.47	-0.409m	2,759.544m	
23	327	0+084.09	0.177m	2,734.257m	
24	404	0+310.39	0.159m	2,741.016m	
25	409	0+273.10	1.693m	2,723.958m	
26	414	0+246.99	-0.737m	2,717.861m	
27	423	0+197.11	-0.207m	2,720.824m	
28	432	0+147.92	-0.782m	2,702.178m	
29	500	0+643.98	-0.245m	2,771.281m	
30	436	0+100.34	0.128m	2,723.870m	
31	443	0+018.10	1.559m	2,761.084m	
32	457	0+491.01	-0.297m	2,769.337m	
33	465	0+516.15	0.653m	2,771.566m	
34	479	0+564.78	0.014m	2,771.098m	
35	481	0+589.89	-0.304m	2,772.217m	
36	489	0+614.20	0.591m	2,770.294m	
37	305	0+042.52	-0.927m	2,756.207m	

Figure E77

As shown in Figure E77, the generated output file will contain the station, offset, and elevation data of every point. This file should be converted into a format suitable for the software to produce a profile from the axis points.

In this format, each station value should be followed by a space and then the corresponding elevation value. A simple example could be:

12.48 1254.63

36.58 1269.67

where 12.48 and 36.58 are the station values and the numbers in front of them are the corresponding elevations.

Note that in this file, the points should be arranged in the ascending order of their station value.

After creating this file from the exported excel file, open the *Profile* menu and select *Create Profile From File* (Figure E78).

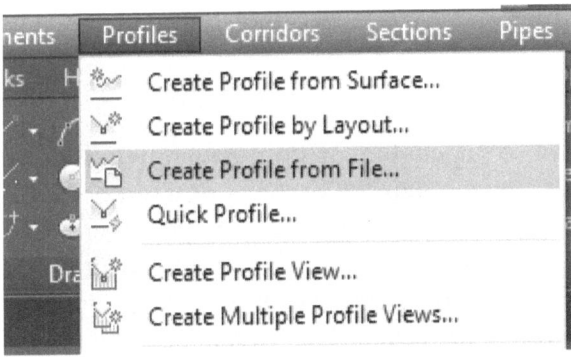

Figure E78

In the opened window, select the created file to open the profile definition window. Adjust the profile display settings as needed and press OK to close the window. Again, open the *Profile* menu, but this time select *Create Profile View* to draw the profile.

68-How to draw the profile of adjacent features alongside a longitudinal profile?

Designers of civil projects are required to maintain accurate information about the elevation condition of nearby features such as pipelines, roads, and rivers. For example, consider a road segment that passes by a river. It is self-evident that to avoid flooding without constructing a retaining wall, the surface of the road segment should be consistently above the river surface. Thus, the profile of the river should also be included in the road profile. This can be done by projecting the features on the longitudinal profile as explained in Exercise 6. The only issue is that if the river path is surveyed separately, the resulting points should be converted into 3dpolyline before being projected on the profile, but if the river path is drawn two dimensionally on a surface, it should be converted into a feature line and the elevations should be derived from the surface during the feature line definition process.

Figure E79

For this exercise, open the 3DPoly-River.dwg file from the folder Project File. This file contains the drawing of a route and its profile, and a river that passes by the route between the stations 00+010 and 01+170. This river is plotted with 3DPolyline. The goal is draw the river profile alongside the route profile. To do this, select the drawn profile, then press the *Project Object To Profile View* button in the ribbon (Figure E79). Then, click on the 3DPolyline of interest and press enter. After pressing enter, the elevation condition of the river will be displayed on the profile as shown in Figure E80.

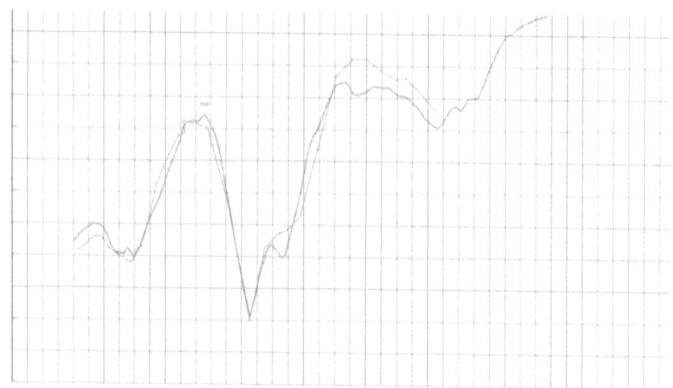

Figure E80

If the feature of interest is not drawn three dimensionally, that is with 3DPolyline, you must convert it into a feature line and instruct the software to derive the elevations from the elevation of the related surface.

The file of this exercise also contains a polyline between the stations 1+950 and 2+910, which represents a gas pipeline.

To convert this polyline into a feature line, open the *Grading* menu, select *Create Feature Lines from Objects*, click on the polyline, and press enter to open the window shown in Figure E31 (as in Exercise 6).

In this window, make sure to check the *Assign elevations* before closing the window. In the next window, specify which surface the elevations should be extracted from, and then press OK to convert the pipeline's polyline into a three dimensional feature line. Now, project this feature line on the longitudinal profile in the same way as explained above.

69-How to draw a band for displaying cross-section, incremental cut/fill volume, and cumulative cut/fill volume under the longitudinal profile?

After designing a route and its project line and calculating the cut and fill volumes, it is recommended to create a band under the longitudinal profile for displaying the cross-sections and cut and fill volumes.

For this exercise, open the file named Volume-on-Band.dwg from the folder Project File. This file contains a surface, an alignment, longitudinal and transverse profiles, and a volume table. The goal is to draw two bands, one for the cut volume and another for the fill volume. The fill band must display the fill operation cross-sections, the fill volume between each two sections, and the cumulative fill volume. Similarly, the cut band must show the cut operation cross-sections, the cut volume between each two sections, and the cumulative cut volume (Figure E81)

45. 23	660. 57 5106. 22	20. 75	274. 18 5380. 41	6. 65	66. 49 5446. 89	0. 00	0. 00 5446. 89	0. 00	0. 00 5446. 89	0. 00	0. 00 5446. 89
0. 00	0. 00 2011. 31	0. 00	0. 00 2011. 31	14. 46	144. 59 2155. 89	31. 97	465. 75 2621. 64	123. 07	1569. 58 4191. 23		1297. 51 5488. 73

Figure E81

In Figure E81, the upper band presents the cut data and the bottom band is related to the fill data. Presented at each section are cut and fill cross-sectional areas and between them are the incremental and cumulative cut volumes above and the incremental and cumulative fill volume below the centerline (the cumulative cut/fill volume presented at each point is the sum of all cut/fill volumes up to that point). Now, we are going to explain how these twin bands can be drawn.

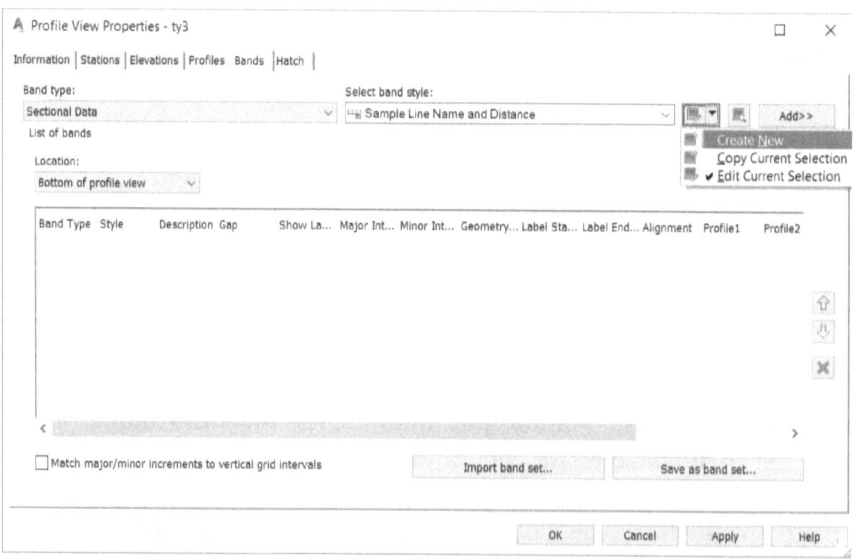

Figure E82

First, select the longitudinal profile, right-click on it and select *Profile View Properties*. In the opened window (Figure E82), go to the Band tab and set the *Band Type* to *Sectional Data*, then open the cascading menu in front of *Select band style* and select *Create new* to start creating a band for displaying the cut operation data. This opens the window shown in Figure E83.

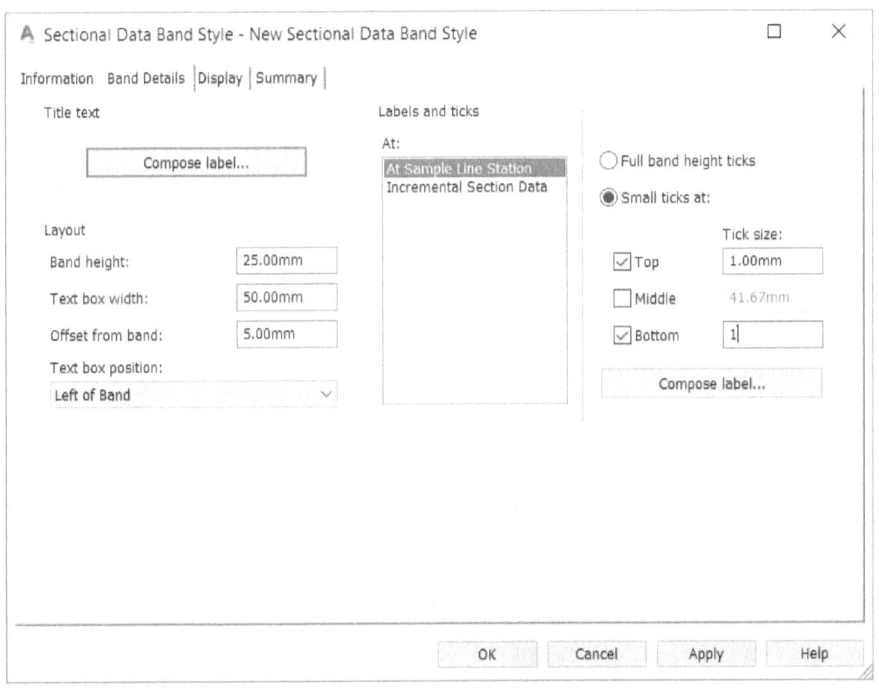

Figure E83

In this window, first open the *Information* tab and type a name for the band style to be created (here, we named it CUT DATA). Then open the *Band Details* tab.

Please note that the labels and ticks section of this tab contains two categories of data. We will use the *At Sample Line Station* option for inserting cross-sectional areas and use the *Incremental Section Data* option for inserting the section volumes. For now, select the *At Sample Line Station* option and then press the *Compose label* button to start inserting the cross-sectional areas. Pressing this button opens the window shown in Figure E84.

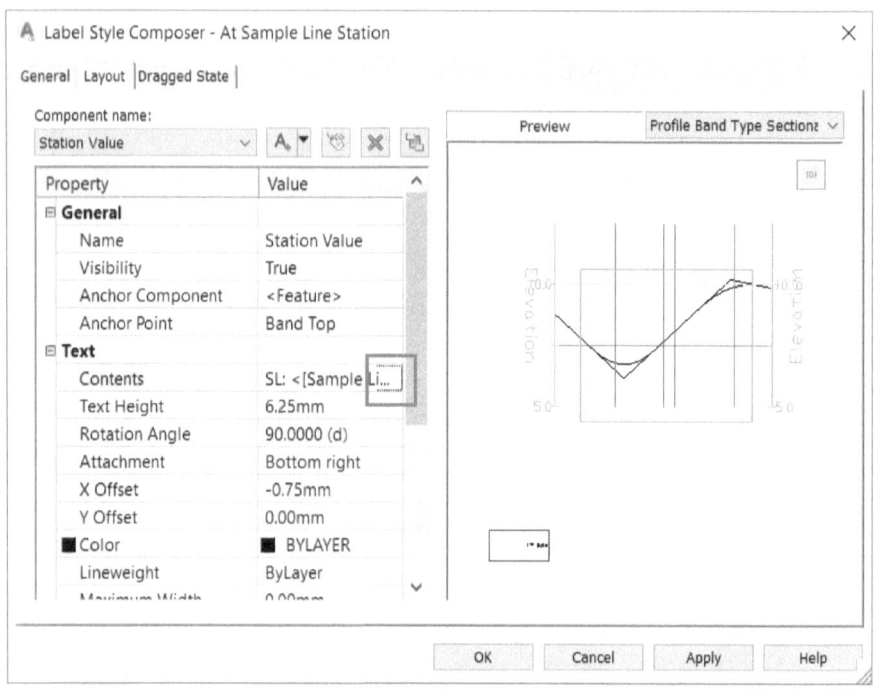

Figure E84

In this window, type a name in the *Name* field, and then click on the field in front of *Contents* to open the window displayed in Figure E85. In the opened window, remove the contents of the right pane, open the *Properties* menu and select *Cut Area at Station*, Press the arrow button to add it to the right pane, and press OK to close this window and return to the window of Figure E84.

In this window, set the *Anchor Point* and *Attachment* menus to *Band Middle* and *Middle Center* respectively so that cross sectional area text appear right between the ticks of upper and lower bands. When done, press OK to close this window and return to the window of Figure E83.

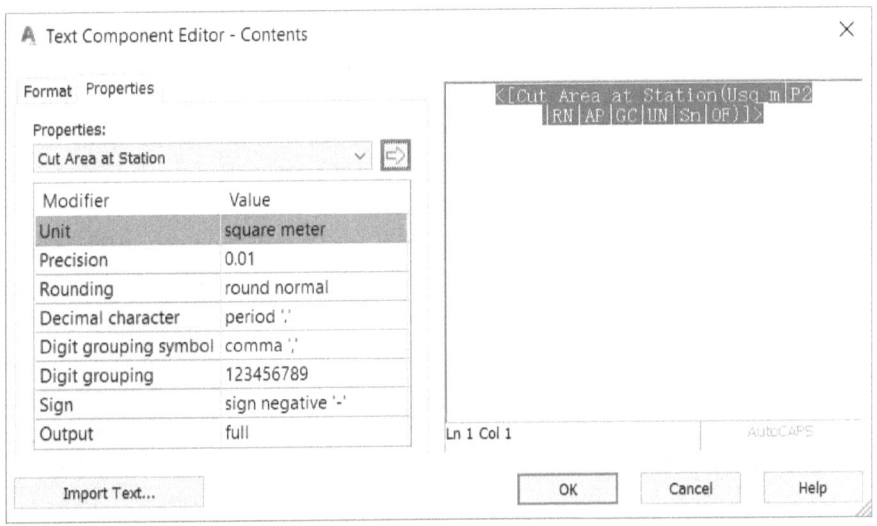

Figure E85

In the window of Figure E83, click on the *Incremental Section Data* option and then press the *Compose label* button to open a window similar to the one shown in Figure E84. As before, type a name for the component and then click on the *Contents* field to start defining the contents of the component. Clicking on this field opens the window shown in Figure E86.

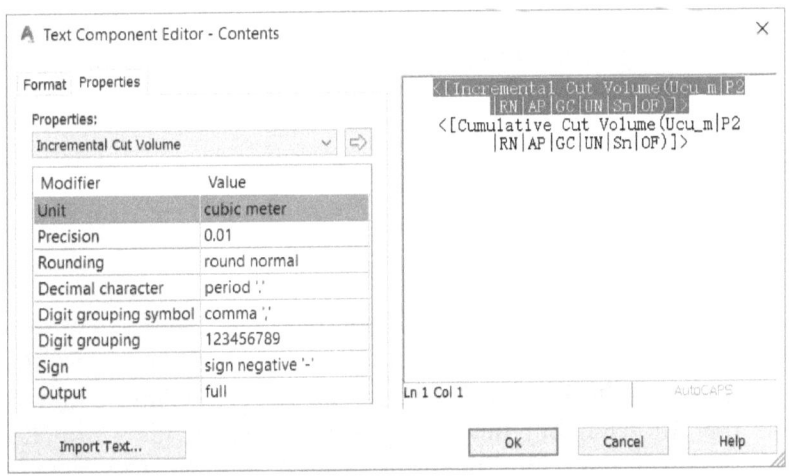

Figure E86

As before, you must delete the contents of the right pane. But this time you are going to insert two types of information, firstly, the cut volume between two sections and, secondly, the cumulative cut volume up to the current section. To do this, open the *Properties* menu and select *Incremental Cut Volume*, and press the arrow button to add it to the right pane. Press enter to go to the next line and then add the *Cumulative Cut Volume* from the *Properties* menu in the same way. After adjusting the settings, including the number of decimal places,..., press OK to return to the previous window. In this window, set the *Anchor Point* and *Attachment* menus to *Segment Mid-Band Middle* and *Middle Center* respectively so that incremental and cumulative cut volumes appear exactly in the middle of the segment. Finally, close this window and the next window by pressing OK to reach the window shown in Figure E87.

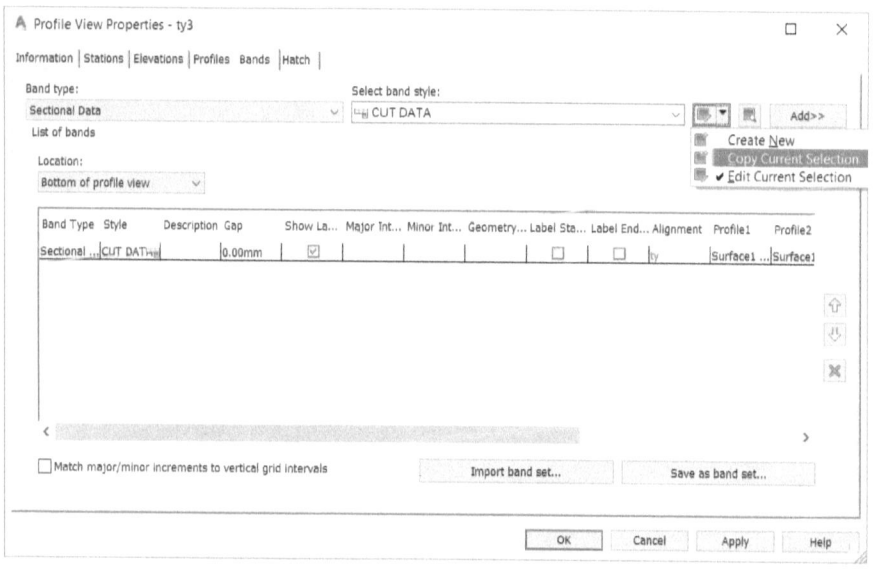

Figure E87

In this window, open the *Select band style* menu and select the band style you just created (we named it *CUT DATA*) and then press the *Add* button to add it to the list. In this step, you have to create another band style for the fill data, but the easier way is to use the *Copy Current Selection* option in cascading menu to make a copy of the created style, and start editing it. When editing this band style, you have to replace the *Cut Area at Station* property with *Fill Area at Station* and replace the

Incremental Cut Volume and Cumulative Cut Volume properties with Incremental Fill Volume and Cumulative Fill Volume, respectively. After making these changes, add the second band style to the list and press OK to see the bands appear as shown in Figure E81.

70-How to produce cross sections for the stations listed in a text file?

In many projects, engineers have to produce cross sections for a series of predetermined stations. For example, suppose that you have conducted an initial survey and produced cross sections at a number of arbitrary stations. Now, after a second survey, you have to produce the cross sections at the same stations to make a match or compare them with each other.

As long as the number of sections is low, this can be done manually. For this purpose, open the *Section* menu and select *Create Sample Lines*, then select the alignment and press enter. In the next window (Figure E88), set the *Sample Line creation methods* menu to *At Station* and then use the command line to enter the stations and their left and right offsets in sequence.

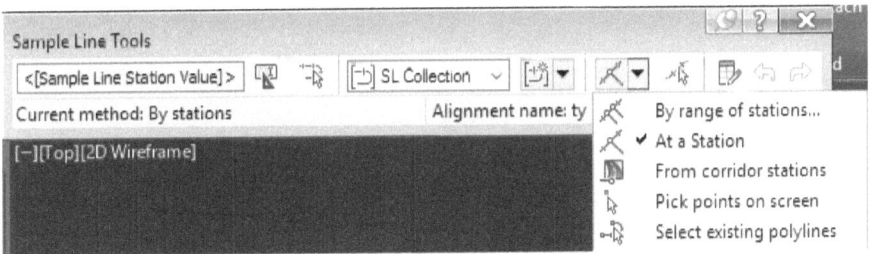

Figure E88

But, when the number of cross-sections is high, this method will not only be time consuming but also very error-prone as numbers should be entered manually.

Civil3D does not provide a specific solution for this issue. Also, there is no way to overcome this problem by the tools provided within the software. But the issue can be easily resolved by writing a simple *Concatenate* formula in Excel.

	A	B	C	D
1	KM	Left Offset	Right Offset	
2	356.37	50	50	
3	412.69	50	50	
4	523.32	50	50	
5	529.65	50	50	
6	534.67	50	50	
7	600	50	50	

Figure E89

For this purpose, open the list of stations in Excel. Then, insert two columns in front of the station column for the left and right offsets (Figure E89). If you have a list of left and right offsets (for example the width of initial cross sections), you can use them in these columns, otherwise, type a constant value for all offsets (here, we set the offset value to 50).

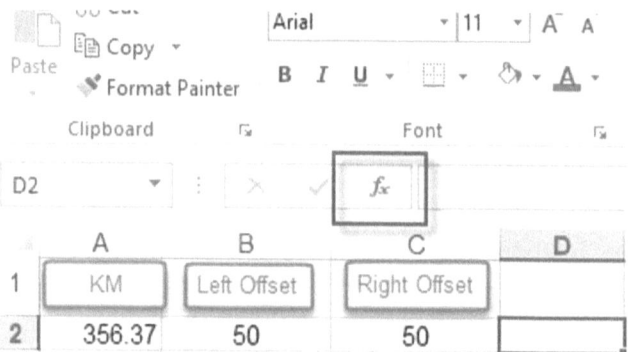

Figure E90

As shown in Figure E90, click on the cell D2 and press the f_x button to open the *Insert function* window shown in Figure E91.

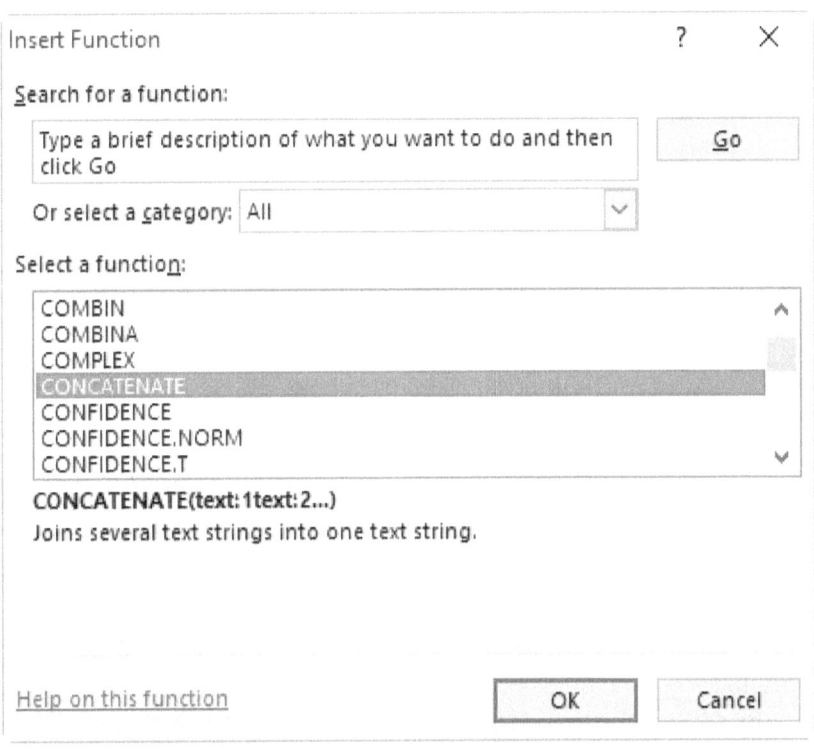

Figure E91

In this window, set the *Or select a category* menu to *All* and find the *CONCATENATE* function from the function list. Pressing OK will open the window displayed in Figure E92, where you must define the details of the selected function.

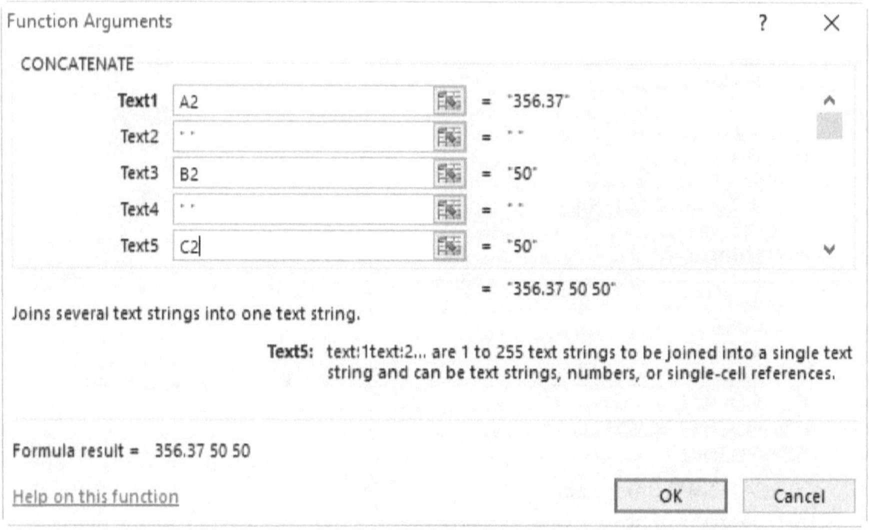

Figure E92

Here, you have to emulate the procedure that must be followed in CIVIL3d to produce cross section at specific stations (i.e. typing the station and pressing enter, typing the left offset and pressing enter, and typing the right offset and pressing enter). For this purpose, press the *Text1* button and click on the first station (here the cell A2), type a space in the *Text2* field to emulate pressing the enter, press the *Text3* button and select the first left offset (here the cell B2), type a space in the *Text4* field to emulate pressing the enter, and finally press the *Text5* button and select the first right offset (here the cell C2). When done, press OK to close the window. The next step is to click and drag down the cell containing the created function (here the cell D2) to make the software reproduce it for the all the rows below (Figure E93). Copy the contents of this column and return to the CIVIL3d environment.

	A	B	C	D
1	KM	Left Offset	Right Offset	
2	356.37	50	50	356.37 50 50
3	412.69	50	50	412.69 50 50
4	523.32	50	50	523.32 50 50
5	529.65	50	50	529.65 50 50
6	534.67	50	50	534.67 50 50
7	600	50	50	600 50 50
8	601.5	50	50	601.5 50 50
9	603	50	50	603 50 50
10	604 5	50	50	604 5 50 50

Figure E93

In CIVIL3d, follow the procedure explained earlier for manual addition of cross sections until you reach the window of Figure E88. In this window, select the *At Station* option and then paste all that was copied from the Excel file into the command line. This result in cross sections being applied to the specified stations. Finally, use one of the routine methods to create cross sections.

71-How to produce cross sections at stations where there is a grade change in the natural ground profile?

Accurate estimation of cut and fill volumes of a project depends on the accurate survey of topographic features followed by the accurate analysis of cross sections. Obviously, it is imperative to assess the cross section at every point where there is a grade change in the longitudinal profile, and failing to do so, even at one point, brings about significant estimation error by neglecting the work that must be done to deal with that grade change before connecting the preceding and succeeding points together.

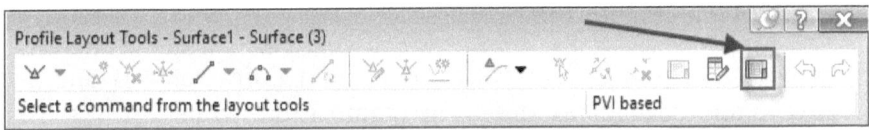

Figure E94

You may know that the software SDR Map provides a feather for creating cross-sections at the stations where there is a grade change. Unfortunately, this task cannot be done with Civil3d alone, so users have to utilize AutoCAD and Excel for this purpose.

The process is similar to what was explained in the previous exercise. In other words, the main challenge here is how to create a list of profile break points in Excel, because after obtaining this list, you just have to follow the steps of the previous exercise to create cross sections.

No.	PVI Station	PVI Elevation	Grade In	Grade Out	A (Grade Change)	Profile Curve Type
1	0+000.00m	2724.427m		9.82%		
2	0+010.92m	2725.500m	9.82%	16.82%	7.00%	
3	0+016.56m	2726.448m	16.82%	8.60%		
4	0+035.37m	2728.066m	8.60%	6.43%		PVI Station
5	0+063.31m	2729.862m	6.43%	5.33%		PVI Elevation
6	0+065.86m	2729.998m	5.33%	-2.16%		Grade In
7	0+071.97m	2729.866m	-2.16%	-1.21%		Grade Out
8	0+084.37m	2729.716m	-1.21%	-7.25%		A (Grade Change)
9	0+093.35m	2729.066m	-7.25%	-6.05%		Profile Curve Type
10	0+097.69m	2728.803m	-6.05%	-13.22%		Sub-Entity Type
11	0+099.66m	2728.543m	-13.22%	-17.69%		Profile Curve Length
12	0+108.36m	2727.005m	-17.69%	-24.70%		K Value
13	0+119.36m	2724.287m	-24.70%	-20.90%		Curve Radius
14	0+126.35m	2722.826m	-20.90%	-12.76%		Asymmetric Length 1
15	0+138.51m	2721.275m	-12.76%	-6.50%		Asymmetric Length 2
16	0+141.83m	2721.059m	-6.50%	-0.88%		Lock
17	0+155.13m	2720.942m	-0.88%	-2.06%		
18	0+167.00m	2720.697m	-2.06%	21.16%		Copy All
19	0+175.61m	2722.520m	21.16%	-5.14%		Copy Selected
20	0+180.77m	2722.254m	-5.14%	-13.54%		
21	0+189.34m	2721.094m	-13.54%	-22.92%		Customize Columns...
22	0+195.85m	2719.601m	-22.92%	12.23%		
23	0+200.54m	2720.174m	12.23%	19.57%		
24	0+218.80m	2723.748m	19.57%	27.91%		
25	0+241.78m	2730.161m	27.91%	23.30%		
26	0+245.19m	2730.955m	23.30%	20.49%		
27	0+250.08m	2731.957m	20.49%	20.79%	0.30%	
28	0+261.57m	2734.347m	20.79%	20.79%	0.00%	
29	0+262.22m	2734.482m	20.79%	20.79%	0.00%	
30	0+262.87m	2734.616m	20.79%	20.79%	0.00%	
31	0+263.51m	2734.750m	20.79%	20.80%	0.00%	

Figure E95

To create this list, click on the longitudinal profile line and press the *Geometry Editor* button on the ribbon. In the opened window, click on *Profile Grid View* (Figure E94).

This will give you a list of all intersection point stations and their respective grade changes as shown in Figure E95. To export this list to Excel, right-click on any row and select *Copy All* and then paste them into a blank Excel file.

The next step must be taken in Excel, where you must convert the station data to numbers. For this purpose, select the column of stations, press Ctrl+H, and delete the characters *m* and + as explained in previous exercises. The rest of the process is similar to what was described in the previous exercise.

Note: The accuracy of cut and fill volume estimations depends on the grade changes not only at the project axis but also at offsets (a longitudinal profile that seems smooth at axis may have significant grade changes along offsets). Thus, to improve the precision of these estimations, it is recommended to produce longitudinal profiles along several offsets (the method is explained in Exercise 1 of this chapter) and follow the above steps for the offset profiles as well.

72-How to produce cross sections at stations where there is a grade change in the project profile?

Accurate production of longitudinal profiles and careful analysis of these profiles in places where there are grade changes in natural ground surface, although extremely important, do not guarantee the accuracy of cut and fill volume estimations. This is because in many projects, and especially in road construction projects, construction work progresses according to a design profile and it is not unlikely to have grade changes in the design profile without any change in the longitudinal profile of the natural ground surface. Failure to produce cross section at the stations where there is a grade change in the design profile means failure to consider the implications of these changes, which is certain to reduce the accuracy of cut and fill volume estimations.

The steps to be followed in this exercise are exactly the same as in the previous exercise, except that you must export the list of intersection points from the design longitudinal profile.

Note: It is recommended to first export the intersection points of the natural ground profile, offsets, and the design profile to Excel, combine them together, sort them in the ascending order of their stations, and then create all cross section at once.

73-How to draw a band for displaying elevation and offset at cross-sectional grade breaks?

When preparing cross sections, it is important to draw a band below the cross section plots for displaying offset and elevation at the points where there is a grade change along transverse direction. An example of this band is illustrated in Figure E96.

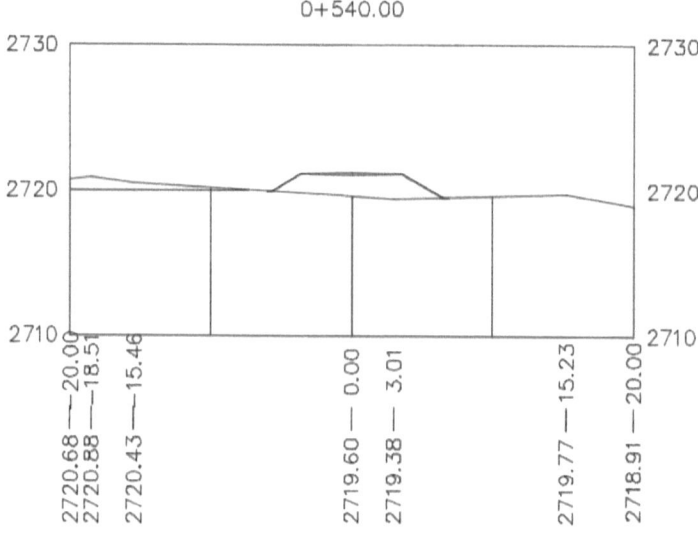

Figure E96

To draw a band like the one shown in the figure, click on one of the created cross section to select it, then right-click on it and select *Section View Group Properties*. Alternatively, you can select one of the sections and press the *View Group Properties* button on the ribbon. Nevertheless, you will be directed to the window displayed in Figure E97.

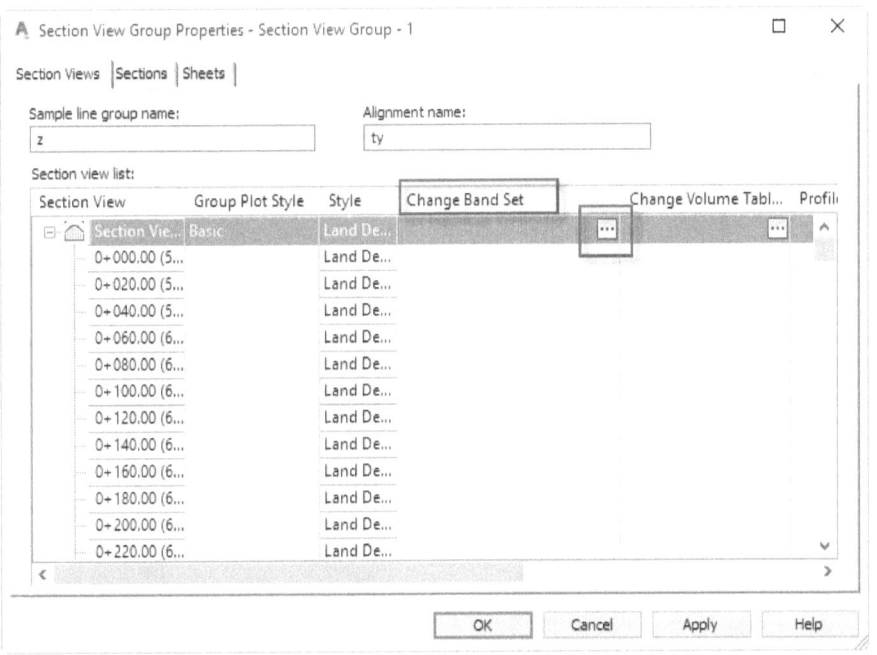

Figure E97

Note: Make sure to select the section itself, not the lines of the cross section

In the opened window (Figure E97), click on the *Change Band Set* button to open the window shown in Figure E98.

In this window, set the *Band type* to *Section Data*, then open the *Select band Style* menu and select *Offsets horizontal*. Then, open the cascading menu to the right and select *Create New* (see Figure E98).

This opens a window, where you have to specify the information that the band must display.

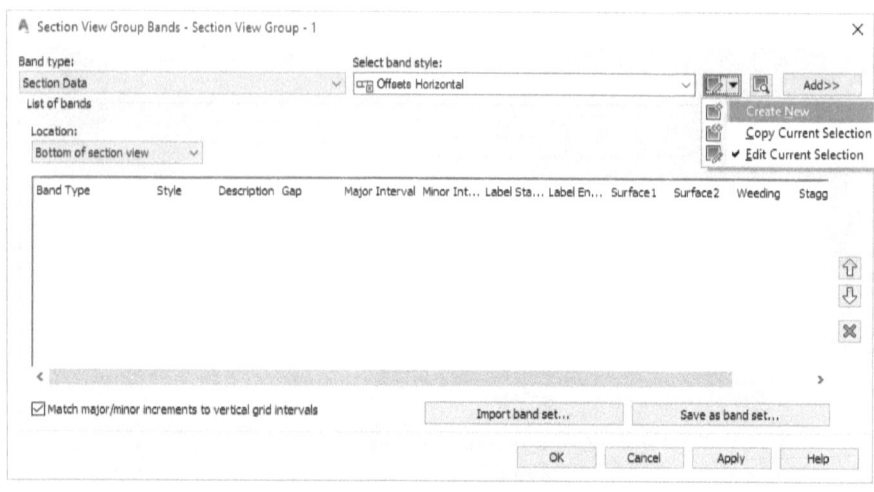

Figure E98

Figure E99

In this window, first go to the *Information* tab and type a name for the band type to be created (here, we named it *offset & elevation*). Next, go to the *Display* tab. Since the goal is to display offset and elevation only at cross-sectional grade breaks, you must turn off all layers except *Labels at Grade Breaks* and *Ticks at Grade Break*. Then, click on *Grade Breaks* and press the *Compose label* button (Figure E99) to open the window displayed in Figure E100.

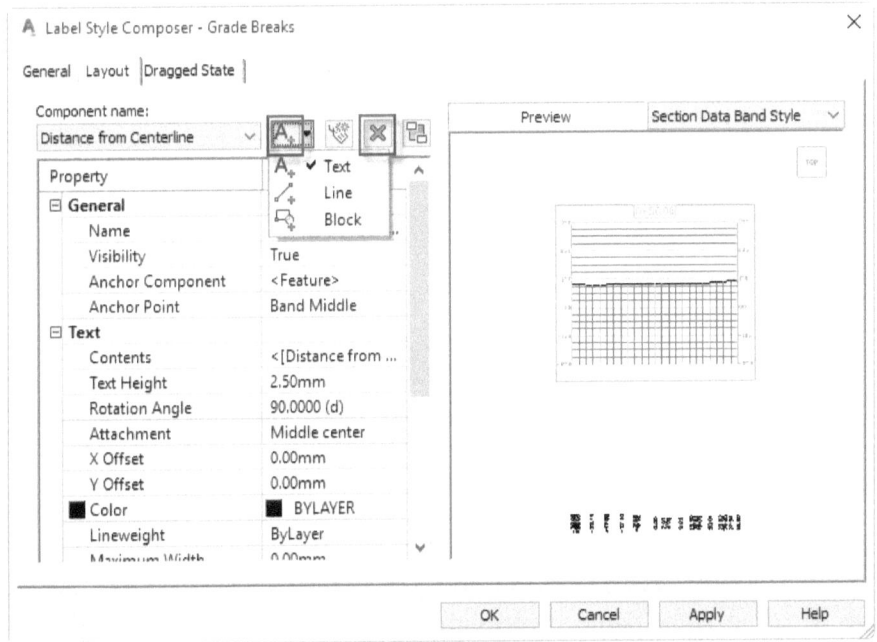

Figure E100

Go to the *Layout* tab of the opened window. In this tab, first remove the current component (with the red X button), then open the cascading menu shown in the figure and select *Text* to create a text-type component for displaying offset. Type a name in the *Name* box (here we named the component *Offset*), then click on the field in front of *Contents* (under the *Text* branch) to open the *Text Component Editor* window displayed in Figure E101.

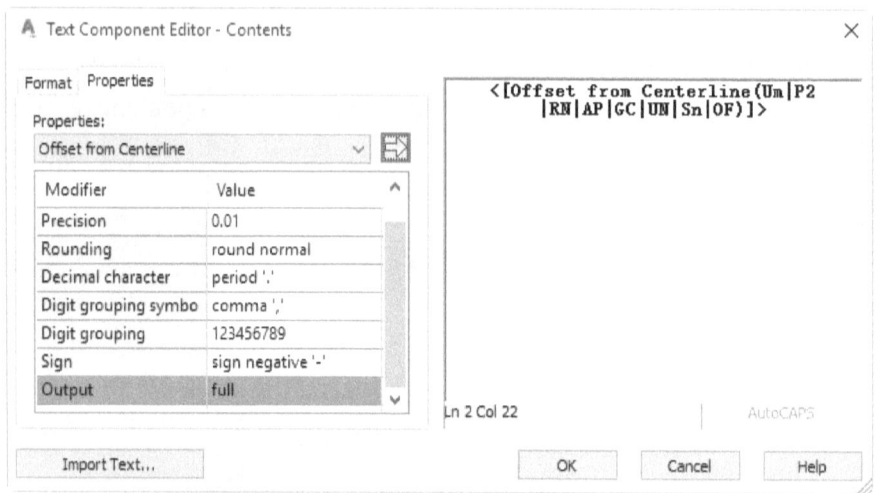

Figure E101

In the same window, you can change *Text Altitude* and *Rotation Angle*. For this exercise, set the text angle to 90 degrees so that texts appear vertically.

In the *Text Component Editor* window, delete the contents of the right pane, open the *Properties* menu and select *Offset from Centerline*, and move it to the right pane by pressing the arrow button. After adjusting other settings, press OK to close the window.

After returning to the window of Figure E100, repeat the above process to create another text-type component, but this time choose the *section1 Elevation* from the *Properties* menu. You must also create a line-type component to add a tick below grade breaks. To make sure that the tick appear exactly below the grade break, set the *Start point at anchor point* to *Band Middle* and set its rotation angle to 90 degree. Adjust the tick size as desired according to the drawing specifications. When done, press OK to close all windows until you return to the window of Figure E98. In this window, select the created band type and press the *Add* button to add it to the list, then close the window by pressing OK. By pressing OK in the next window, you will apply the changes to all cross sections.

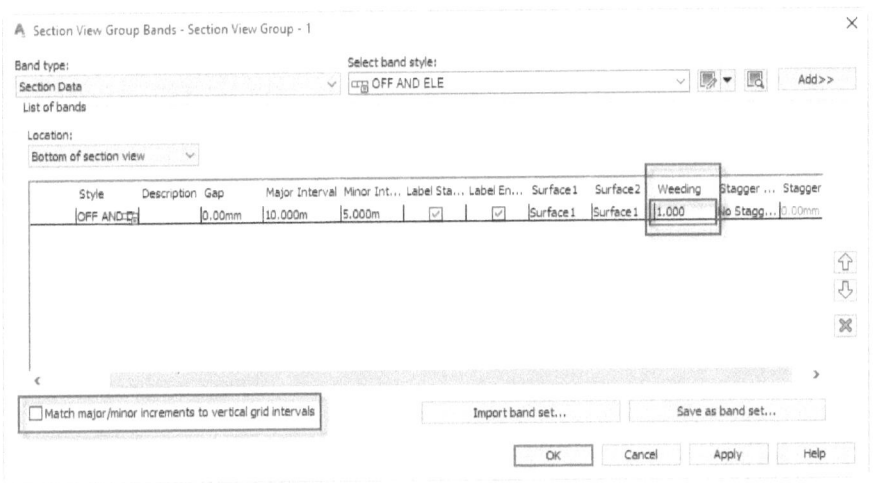

Figure E102

There are two suggestions that you should consider. First, after adding the band type, you should consider unchecking the *Match major/minor* option and setting the *Weeding* threshold to 1 to ignore the break point with the spacing of less than 1 meter (Figure E102).

The second suggestion is that in the window of Figure E100, you should set the *Y Offset* of either elevation or offset component to a value such as -5 in order to prevent elevation and offset texts from overlapping with each other.

74-How to display the cross-section ID number below the station label?

When drawing or working with cross sections, it is convenient to have quick access to the cross section ID numbers alongside station details. By default, Civil3d only displays the station information, so you have to take the following steps to add the cross section ID numbers to the cross section label.

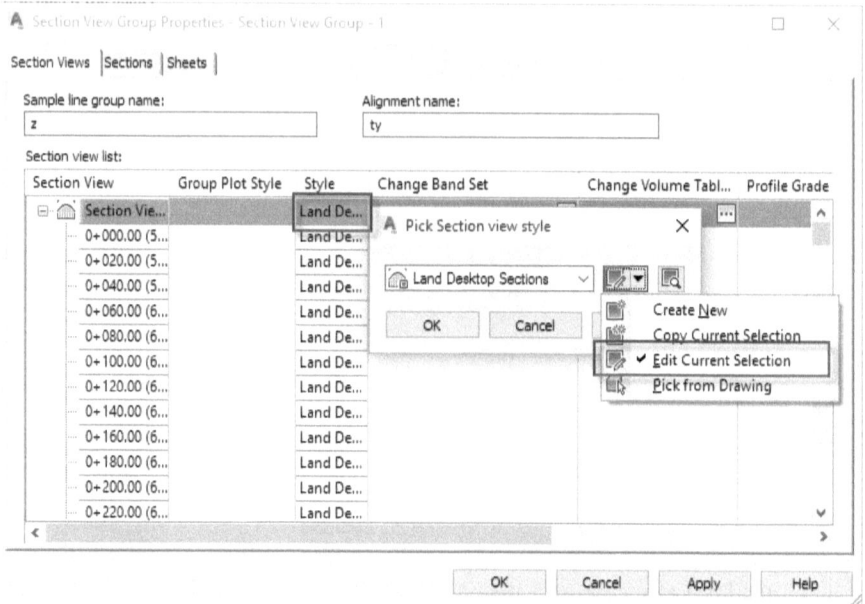

Figure E103

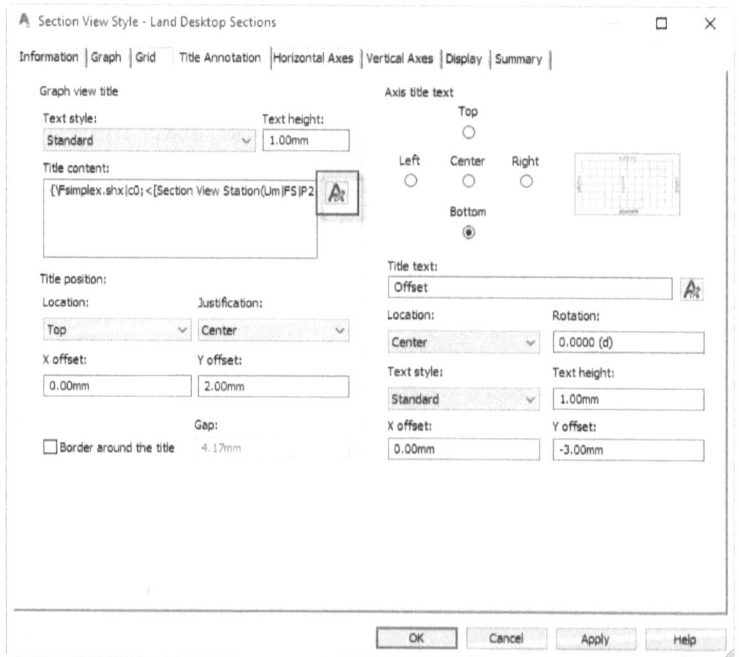

Figure E104

Select one of the sections, then right-click on it and select *Section View Group Properties*. Alternatively, you can select one of the sections and press the *View Group Properties* button on the ribbon. Nevertheless, the software opens the window shown in Figure E103.

As shown in Figure E103, click on one of the styles listed in the *Style* column and select *Edit Current Selection* in the cascading menu to open the window displayed in Figure E104.

In the window, go to the *Title Annotation* tab and click on the button shown in Figure E104 inside the *Graph View Title* pane. This button opens the window shown in Figure E105.

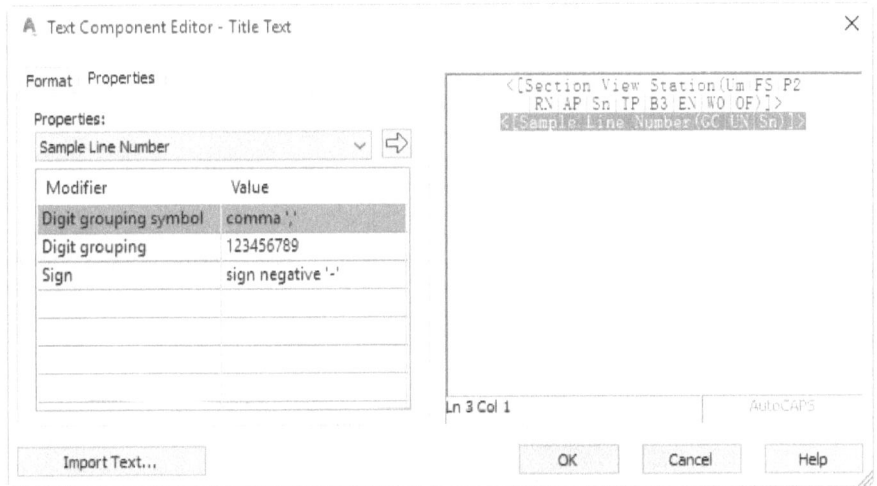

Figure E105

In this window, create a new line under the contents of the right pane, open the *Properties* menu and select *Sample Line Number*, and press the arrow button to move it to the right. The end result should be as shown in Figure E105. When done, press OK to close all windows. The cross-section ID number must now appear below the station information as shown in Figure E106.

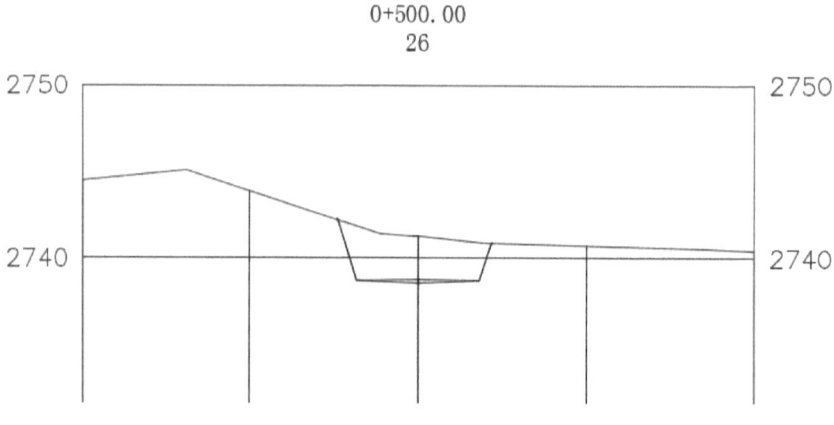

Figure E106

75-How to define a coordinate system and zone number for a drawing file?

As shown in Figure N1, open the *Settings* tab in the *TOOLSPACE* window, right-click on the name of the drawing file and select *Edit Drawing Settings* to open the window shown in Figure N2.

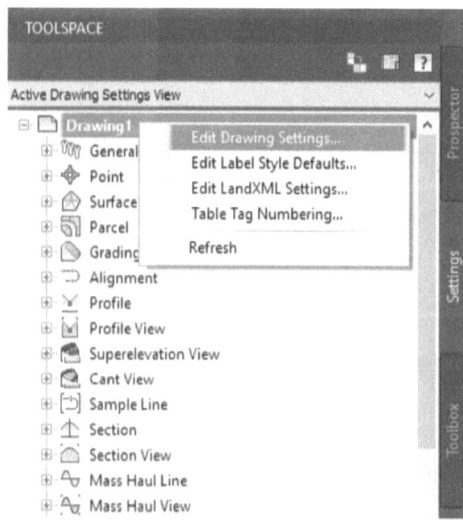

Figure N1

Open the *Unit and Zone* tab of this window. In the *Zone* pane of this tab, in front of *Categories*, you can select the projection and coordinate system, and in the field below *Available Coordinate system* area, you can select the zone number of the area surveyed. When done, press OK to close the window.

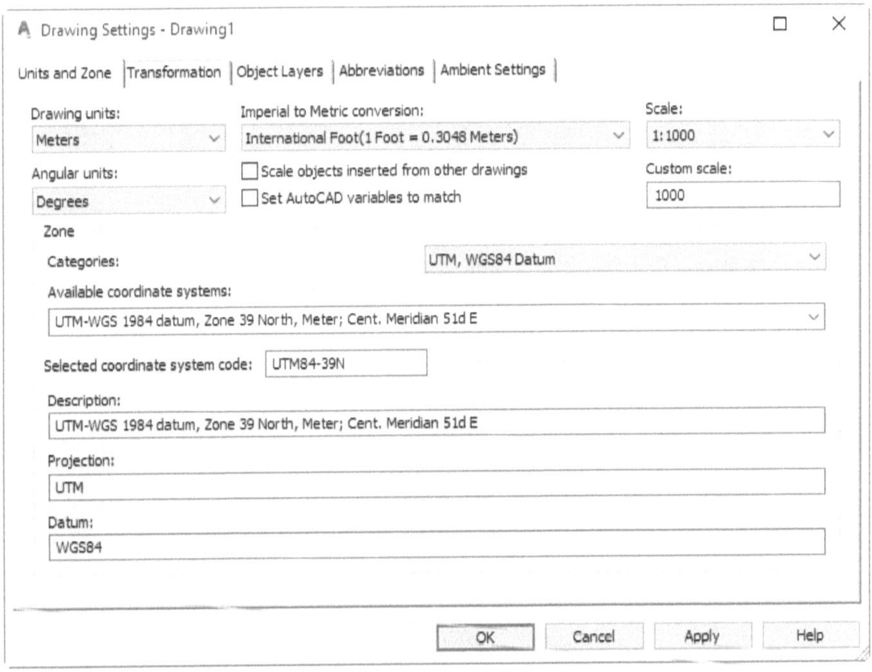

Figure N2

Note: If your drawing file lacks a global coordinate system, you will not be able to access parameters like geodetic latitude and longitude and scale factor.

76-Automatic drawing of lines and surface features

The purpose of this exercise is to help you understand the concept of automatic drawing. After mastering this capability, you will be able to use it at your convenience in busy drawings (such as rural and urban cadastral mapping projects) for automatic drawing of lines and boundaries of features. Consider Figure N3 for example. In this example, a building and a road have been surveyed using the reciprocal method. The building points have been surveyed with the code BLD, the points on one side of the road have been surveyed with the code R1, and those on the other side have been surveyed with the code R2. However, there are several exceptions to this rule.

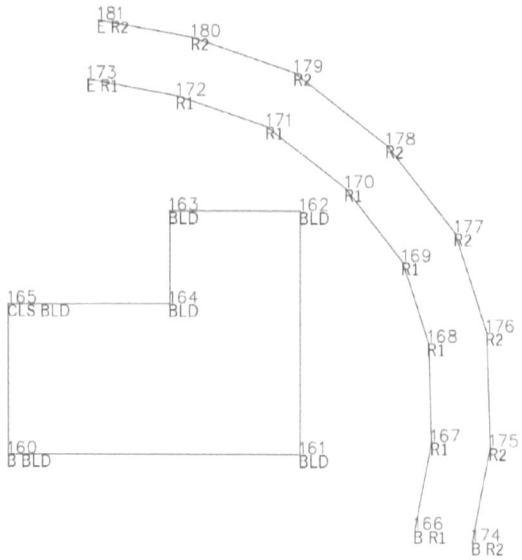

Figure N3

The point 160, which is the first point surveyed in the building, has been given the code BBLD (the letter B stands for Begin), and the point 165, which is the end point that closes the loop to the point 160, has been given the code CLS (CLS stands for Close). Also, the points 166 and 174, which are the first points of the lines R1 and R2 have been given the codes BR1 and BR2, and the points 173 and 181, which are the end points of these lines have been given the codes ER1 and ER2 (the letter E stands for End).

Note: When you survey a series of points like R1 with the aim of using the automatic drawing feature, it is important to respect the order and sequence of points. In Figure N3, for example, you cannot jump from the point 166 (R1 line's beginning) to the point 170 and then return to the point 168. Rather, you must survey the points in the same order in which you want them drawn.

Note: When you want to use a code string like R1 for automatic drawing, you cannot survey the points on both sides of the road simultaneously. Instead, you must first finish working on one side of the road and then proceed to the other side. If you want to survey both sides at the same time, you have to define another code string such as R2 for the other side, and assign the start, middle, and end points of that side with that code.

For this exercise, open the file <u>Auto Drawing.txt</u> from the folder <u>Points</u>. This text file contains the data of the points illustrated in Figure N3. The goal is to import this file into CIVIL 3D and use the automatic drawing capability of this software.

For this purpose, open the *Survey* tab in the *TOOLSPACE* window, right-click on *Equipment Database* and select *New* (Figure N4).

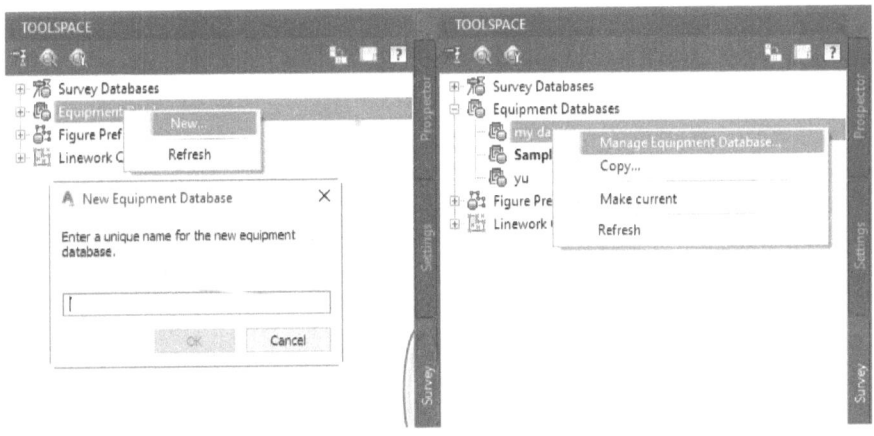

Figure N4

Type a name for your database. Click on the button next to *Equipment Database* to view the existing databases, then right-click on the database you created and select *Manage Equipment Database*. In the opened window, specify the units used for length, angle, etc. and press OK to close the window.

Next, right-click on *Linework Code Sets*, select *New* and type a name for your linework (Figure N5). After confirming the name, you will be directed to the window shown in Figure N6, which provides a list of all operator codes and allows you to edit them. For example, you can change the code that represents the beginning point from B to ST.

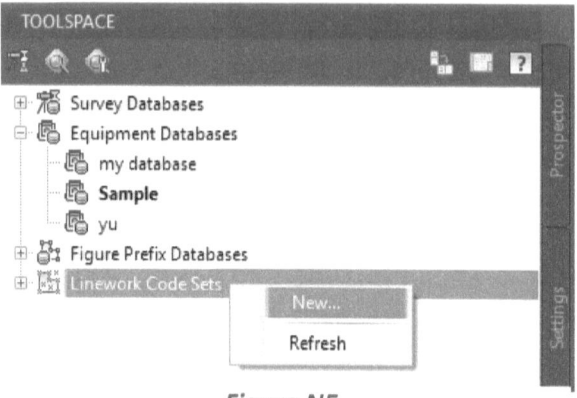

Figure N5

Now you have to create a survey database. To do this, right-click on *Survey Database* and select *New local survey database* (Figure N7). Choose a name for your survey database and press OK. As shown in Figure N8, right-click on the survey database that you just created and select *Edit survey database settings*. In the opened window, adjust the survey database settings, such as the global coordinate system, the zone number of the surveyed area, etc. and then press OK to close the window.

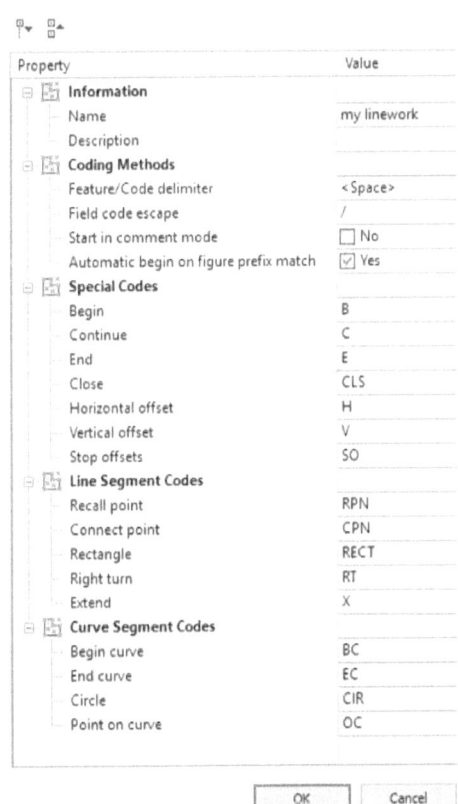

Figure N6

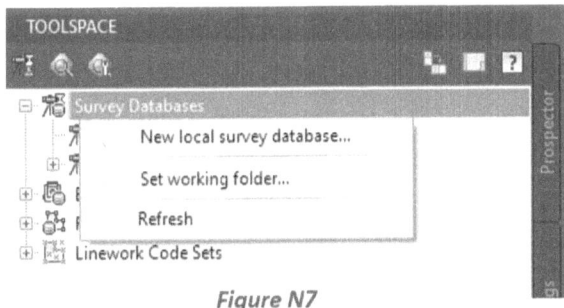

Figure N7

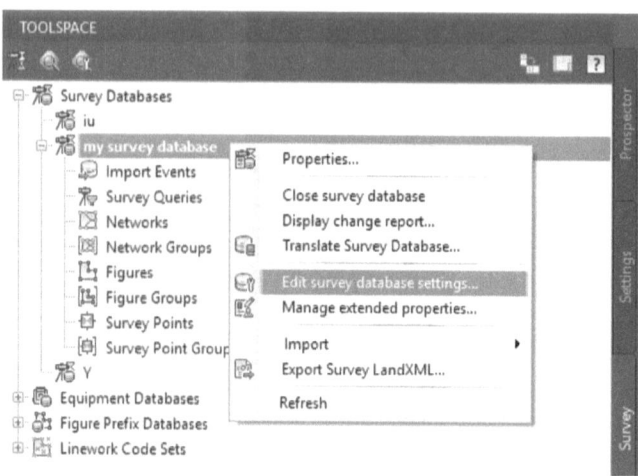

Figure N8

The next step is to import the survey data and initiate the automatic drawing. To do this, click on the + sign next to the name of your survey database to expand it, then right-click on *Import Event* and select *Import Survey Data*. In the first page of the opened window, click on the survey database to select it and press *Next*. In the next page, select the file and specify its format, and then press *Next*. In the opened page, use the *Create New Network* option to create a network for the points to be imported and give it a name, and then go to the next page. In this page, you must adjust the settings of automatic drawing as shown in Figure N9.

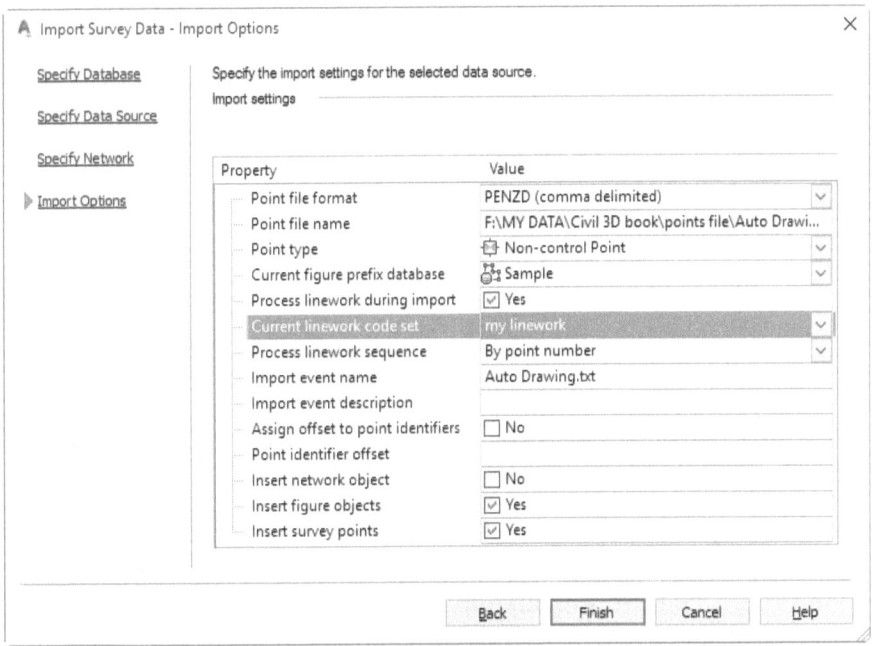

Figure N9

In this window, remember to tick the *Process linework during import* option and set the *Current linework code set* menu to the linework you created. When done, press the *Finish* button to start importing the points and initiate automatic drawing. After pressing this button, the software must start drawing all the lines automatically.

77-How to place an object in parallel with another object?

The *Rotate* command can be used to rotate an object such that one of its sides becomes parallel to a line or a side of another object.

For example, suppose that in the drawing of Figure M1, you want to rotate the triangle ABC about the point A such that the side AB becomes parallel to the line DE.

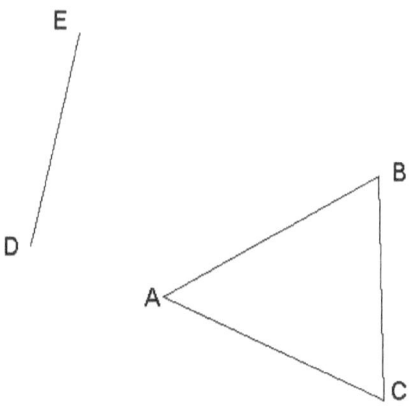

Figure M1

For this purpose, type the shortcut command *RO* in the command line and press enter. Then click on the triangle and press enter. Next, click on the point A as the base of rotation. The next step is to type *R* and press enter (to select the *Reference* option). Then you have to click first on the point A and then on the point B to designate the side AB as the origin side. Next, type *P* and press enter (to select the *Points* option). Then, click first on the point D and then on the point E to designate the line DE as the destination side. After pressing enter, the triangle ABC should rotate such that the side AB becomes parallel to the line DE. This form of the *Rotate* command has extensive use when working with survey drawings.

To practice, open the file Rotate2.dwg file and follow the instructions given below.

Type RO.

- Command: RO

- ROTATE

- Current positive angle in UCS: ANGDIR=counterclockwise ANGBASE=0

In this step, you must click on the target triangle or any object you want rotated and then press enter.

- Select objects: 1 found

- Select objects:

- Specify base point

Here, you must click on point A as the base of rotation.

- Specify rotation angle or [Copy / Reference] <0>: R

Now type *R* to select the *Reference* option and press enter.

- Specify the reference angle <0>:

Click on the point A.

- Specify second point:

Click on the point B.

- Specify the new angle or [Points] <0>: P

Here, you must type P to select the *Points* option and press enter.

- Specify first point:

Click on the point D.

- Specify second point:

Click on the point E.

78-Difference between the commands Layon and Layuniso

Some users think that the commands *Layon* and *Layuniso* do the same thing, but obviously this is a wrong misconception. To understand the difference of these commands, consider the following example.

A drawing contains 20 layers, of which 17 are enabled and 3 are disabled. Now, after enabling one of the enabled layers with the *Layiso* command, we will have 1 enabled layer and 19 disabled layers. Note that of these 19 disabled layers, 3 were disabled from the beginning, and the remaining 16 were disabled by the *Layiso* command. Now, if we use the *Layuniso* command, only the 16 layers that got disabled by the *Layiso* command will be re-enabled. In contrast, using the *Layon* command will enable all the layers, including those three that were originally disabled.

79-Identification of open land plot polylines

When drawing cadastral maps, it is important to ensure that all land plot polylines are closed. This is important because open land plot polylines can cause error in extracting information, making measurements, adding information labels, and linking the maps to ArcGIS databases. For a small drawing, this can be done manually by checking land plot polylines one by one. But for large drawings, manual checking of all land plot polylines will be extremely difficult and time consuming. An easy way to accelerate this process is to use the *Quick Select* command. To learn this method by practice, open the file plineclose.dwg, which is a drawing of 13 land plots.

As you can see, it is difficult to visually distinguish the closed plots from the open ones. To identify the open plots, type and execute the shortcut command *qselect* to open the *Quick Select* window shown in Figure M2. As shown in the figure, set the *Object type* menu to *Polyline* and set the *Properties* menu to *Closed*. Since the goal is to select open polylines, set the *Operator* menu to *<> Not equal* and select the *Yes* option in the *Value* menu. After pressing OK, the polylines with the specified property will be selected.

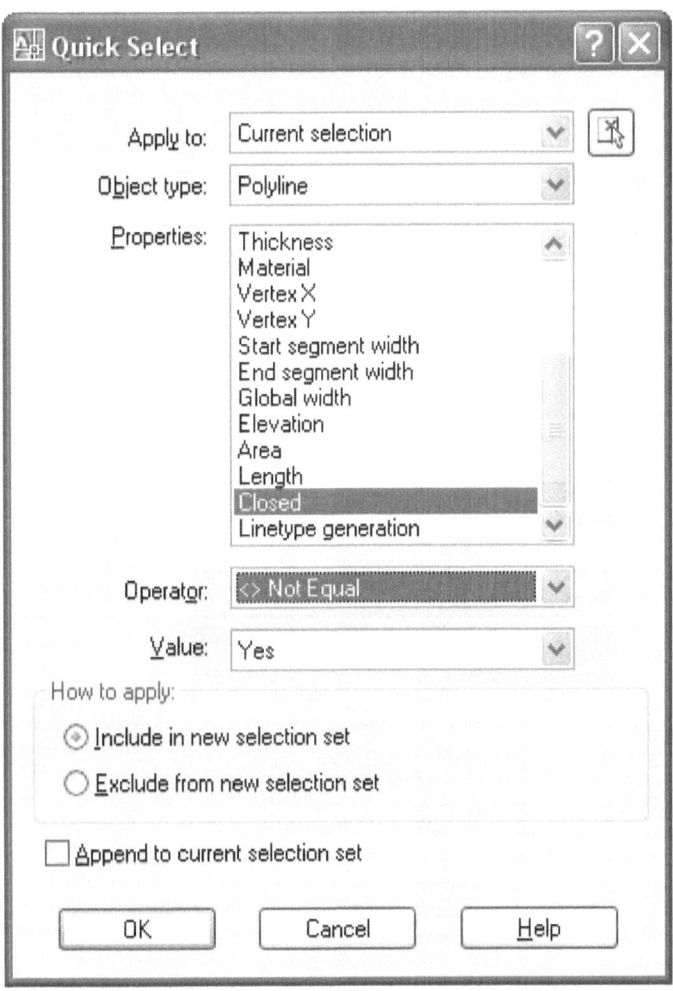

Figure M2

80-Selection of land plots with a specified area

Another convenient use of the *Quick Select* command when drawing cadastral maps is in finding the land plots with a specified area. For example, suppose that we want to create a list of all land plots whose area is greater than a certain threshold. To do this, we need to first find and select those plots. For this exercise, open the file qselectarea.dwg, which is a drawing of 29 land plots with different areas. Here, the goal is to create a file containing only the land plots that are larger than half a hectare (5,000 square meters). To do this, run the *qselect* command in the command line to open the *Quick Select* window. In this window (Figure M3), set the *Object type* menu to *Polyline*, set the *Properties* menu to *Area*, set the *Operator* to *>Greater than*, and set the *Value* to 5000. After pressing OK, the *Quick Select* window will be closed and all land plots with an area of more than 5000 square meters will be selected.

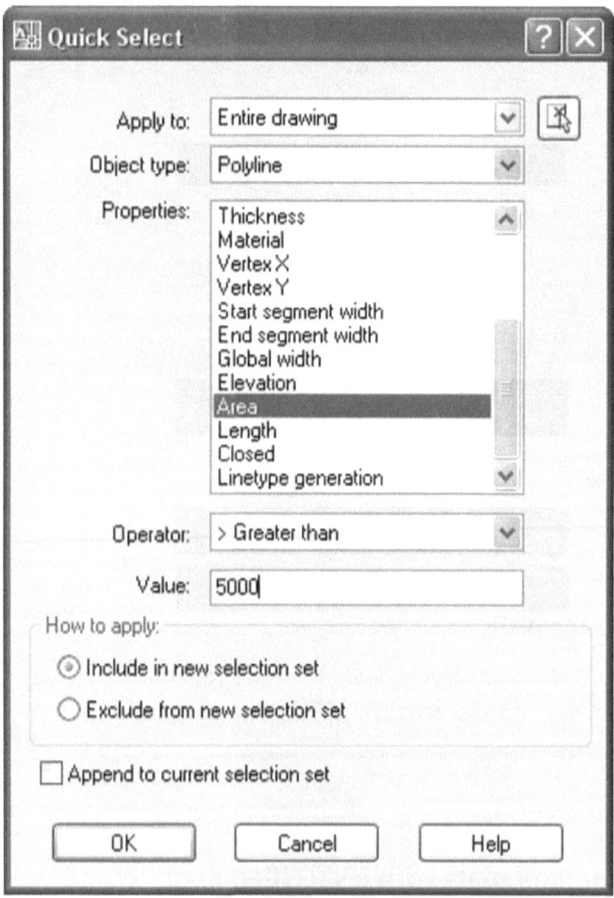

Figure M3

To transfer these plots to another file, simply execute the *Copy* command and paste them in the other file. Further details regarding Copy and Paste operations will be provided in the next chapter.

Please note that while advanced users may find such exercises rudimentary, the importance of this command and such simple solutions cannot be overstated. Moreover, mastering this command can assist all users to find their own convenient solutions for more complex drawing problems.

81-How to disable AutoCAD commands

You can use the Undefine command to disable an AutoCAD command and prevent the software from recognizing it in the future. For example, if you want to disable the *Line* command, type *Undefine* in the command line and press enter, and when the software ask you to *Enter command name*, type *line* in the command line.

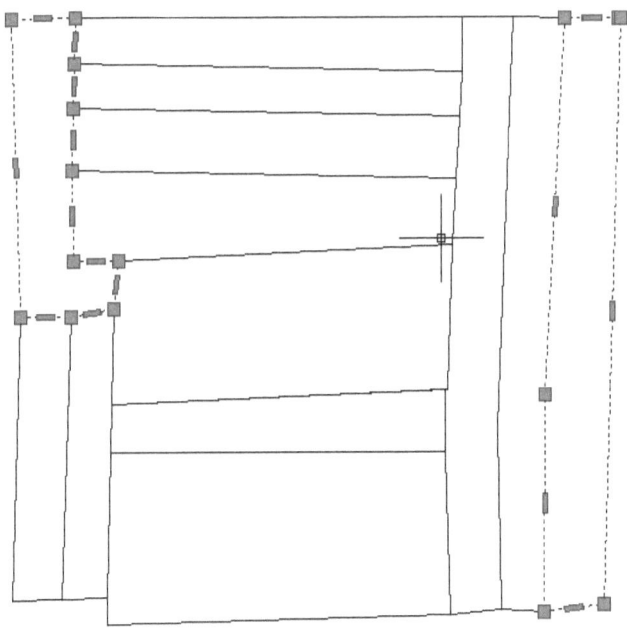

Figure M4

Following this instruction, the software will not allow the *Line* command to be executed through the command line, menu, or toolbars.

You can use the *Redefine* command to enable a disabled command.

As shown in Figure M4, this command will select all open plot polylines, which in this example, include one plot in the upper left and another in the right side of the drawing.

Now, you can change the color of these plots to mark them for editing or further investigation.

82-Automatic repeating of commands for accelerated drawing

Perhaps you know that the command *Multiple Point* can be used to quickly create a large number of points in succession. Similarly, to create a large number of circles in succession, you can type the command *Multiple* in the command line and then add *Circle* to initiate successive execution of the *Circle* command.

Similarly, you can use the *Multiple* command to repeat any command you desire. This command is helpful in situations where a repetitive task must be performed quickly and without interruption.

83-How to modify command shortcuts?

Professional users often execute commands with shortcuts and usually modify these shortcuts to further accelerate the process. For example, a user who has make repeated use of *Layiso* shortcut (to run the *LayerIsolate* command) may prefer to change this shortcut to a much shorter shortcut such as *01*. Indeed, the software can be modified to execute the *Layer Isolate* command whenever the user enters the shortcut 01 in the command line. The method of this modification is explained below.

First, open the *Express* menu and select *Tools* and then *Command Alias Editor* (Figure M5).

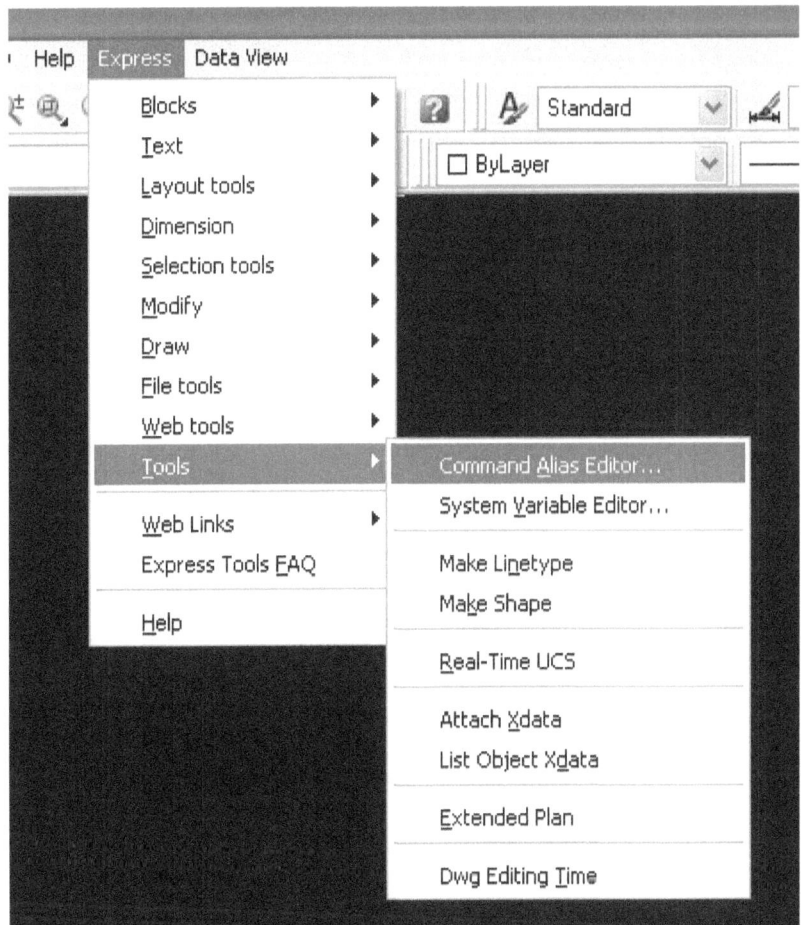

Figure M5

This opens a window named *acad.pgp - AutoCAD Alias Editor*, which is displayed in Figure M6.

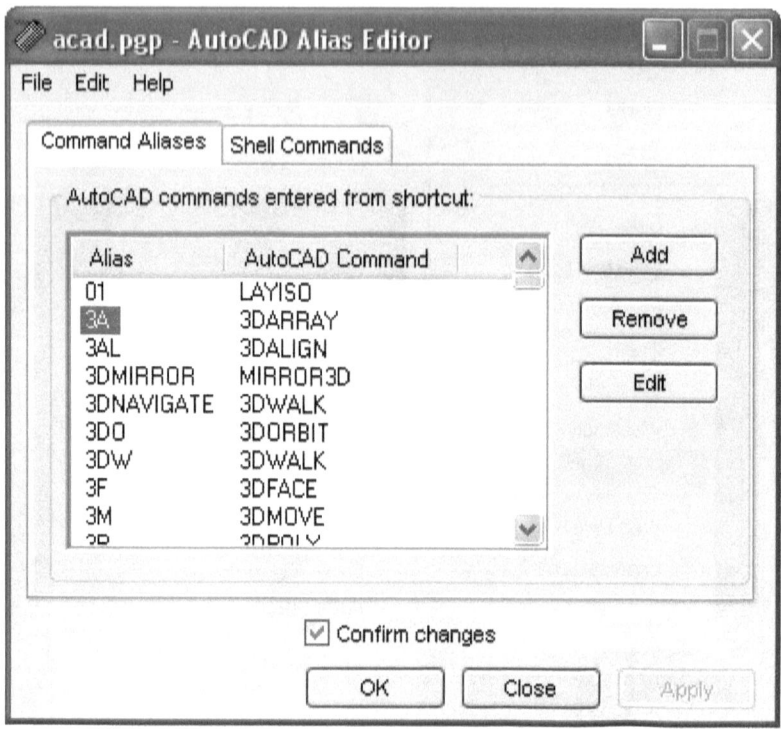

Figure M6

To assign the shortcut *01* to the *Layiso* command, press the *Add* button to open the *New Command Alias* window shown in Figure M7.

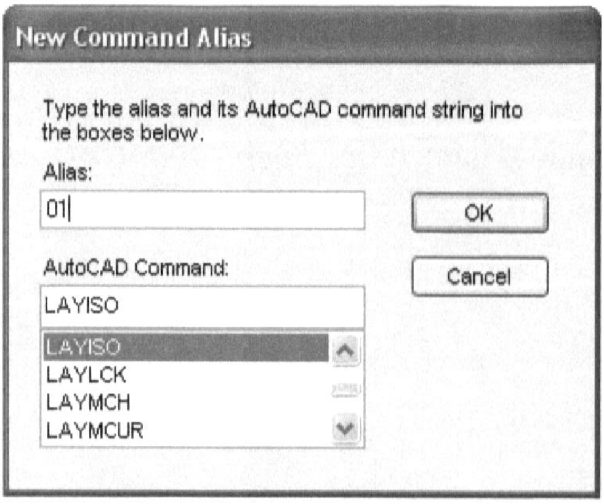

Figure M7

In this window, find the command of interest in the *AutoCAD Command* menu and type the desired shortcut in the *Alias* field. After pressing OK and closing all the windows, you can use the shortcut *01* to run the command *Layiso*.

84-How to make a new Icon?

To make a new Icon, execute the shortcut command *CUI* or open the *Tools* menu and select *Customize* and then select *Interface* to open the *Customize User Interface* window (Figure M8).

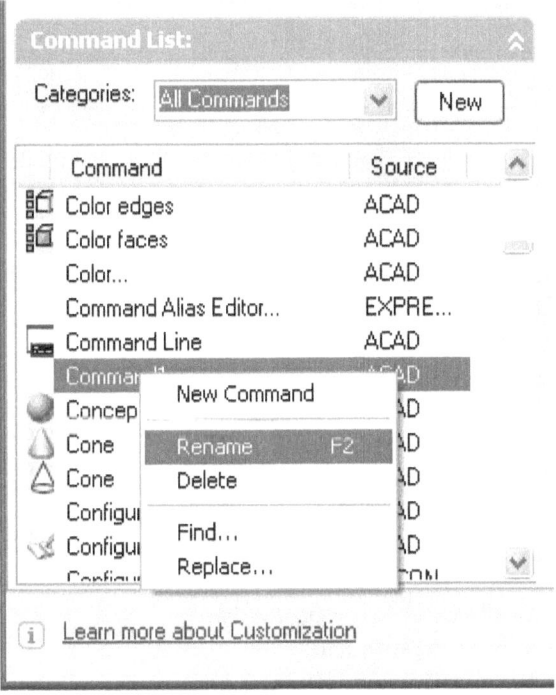

Figure M8

In the lower left part of this window, there is a box named *Command List* and a bottom named *New*, which you must press to create a new command (named *Command1* by default). You can right-click on the created command and select *Rename* to change its name.

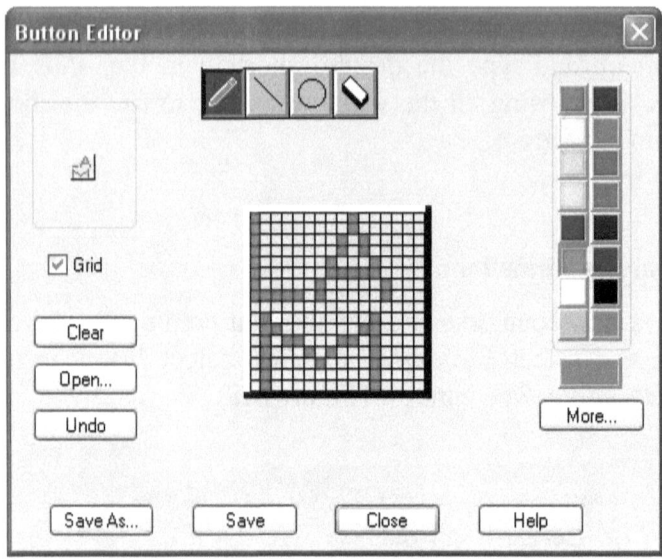

Figure M9

In this example, we named the command *Laymerge* (because we want the Icon to execute the *layermerge* command).

Now click on the command and press the *Edit* button in the *Button Image* section to open the *Button Editor* window shown in Figure M9.

Use the tools provided in this window to design an icon. Once done, press the (*Save*) button to save the icon, and assign it to the created command.

Note that if you click on any command, you will see its properties in a box named *Properties* to the right. For example, if you click on the *Ellipse* command (used for drawing ellipses), you will see the *Properties* window shown in Figure M10.

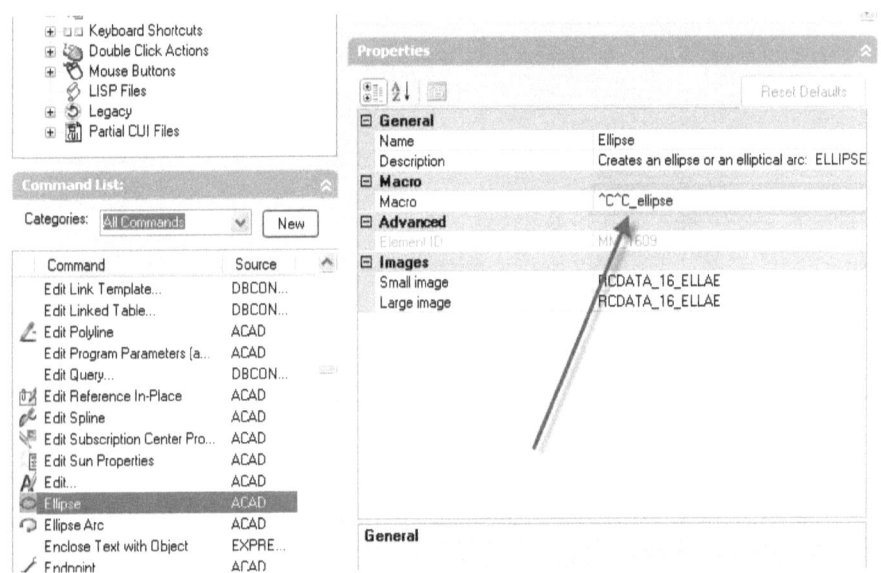

Figure M10

One of these properties is the macro code of the command.

As you can see, the macro code of the *Ellipse* command is ^C^C_ellipse.

If you copy this code into the macro field of the command you just created, it will function exactly the same way as the *Ellipse* command (i.e., it will draw an ellipse).

But since we want the button to execute the *Layermerge* command, we must find this command in the command list, click on it to see its macro code in the *Properties* window (Figure M11), copy this code, and paste it in the macro field of the created command.

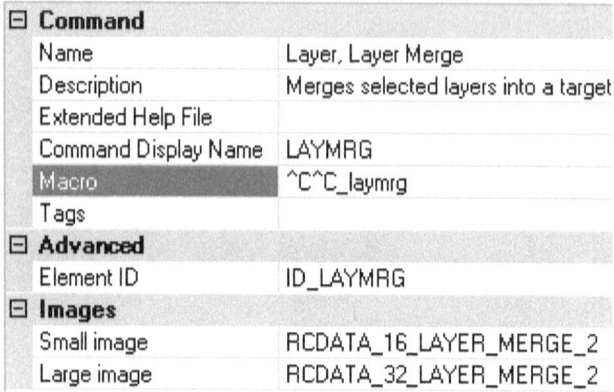

Figure M11

So far, we have created a command that performs the function of the *Layermerge* command.

The next step is to place this command in the *Draw* toolbar after the *Line* command.

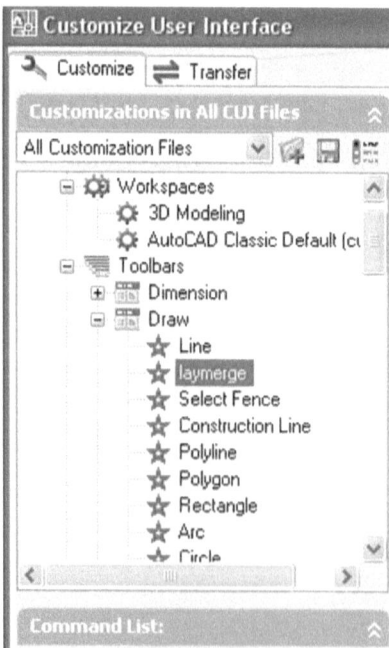

Figure M12

To do this, find a box named *Customizations in All CUI File* in the same window. Expand the *Toolbars* branch and then the *Draw* sub-branch. Find the created command in the box positioned below and drag and drop it below the *Line* command in the *Draw* sub-branch. After pressing OK, you must see the button appear in the defined position in the *Draw* toolbar.

This method will be especially rewarding if you master the macro codes, because then you will be able to create buttons with the codes modified for accelerated completion of repetitive tasks. For example, if you create a button with the macro *^C^C_erase all save A close*, the function of the button will be to delete all the objects in the drawing, save the file with the name A, and then close the AutoCAD.

Figure M13

A more practical example is to create a button with the macro *^C^C_pedit m all y j .01* for converting all *Lines* to *Polylines* and joining them together with just one click.

85-How to create buttons for a LISP or macro?

Before learning how to create buttons for LISPs and macros, you must learn how to load them.

To load a LISP file, open the *Tools* menu and select *Load Application* to open the window shown in Figure M14. Select your LISP file in the *Look in* section, press the Load button, and then close the window.

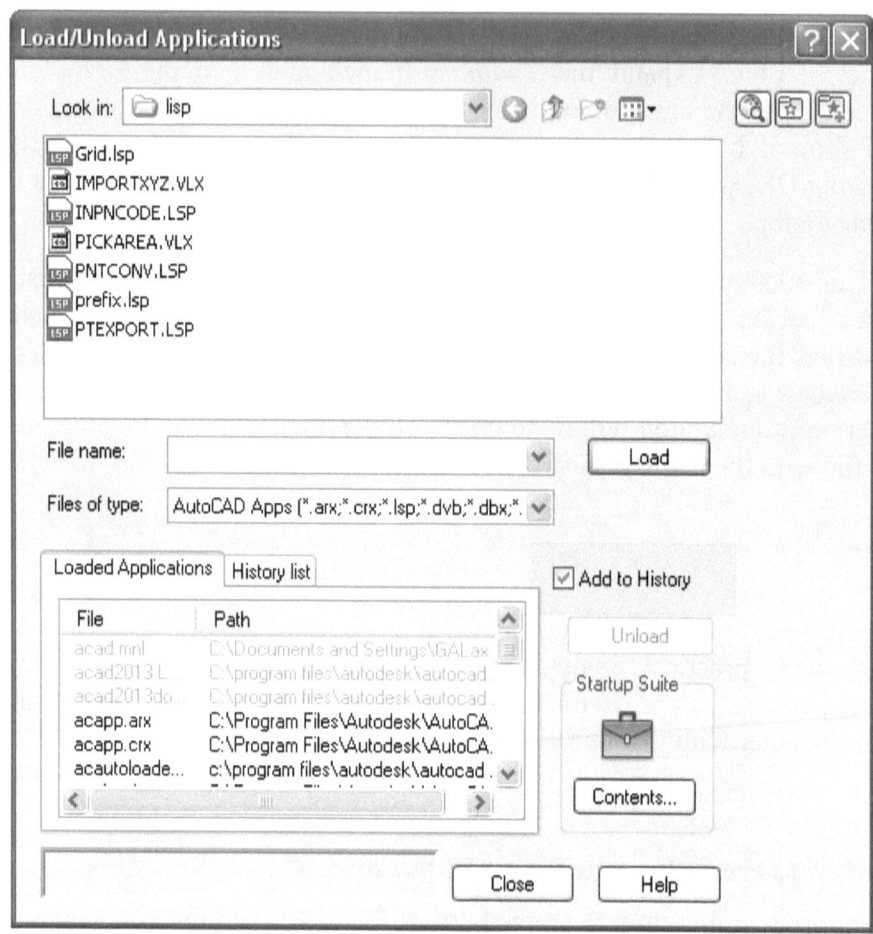

Figure M14

It should be noted that if you tick the *Add to History* checkbox before loading, the file will be permanently stored in AutoCAD, otherwise, you will have to reload the file every time you start the program.

After loading the LISP file, you can run it by typing its shortcut in the command line.

To load a macro file, you must open the *Tools* menu and select Macro and then *VBA Manager* to open the window displayed in Figure M15.

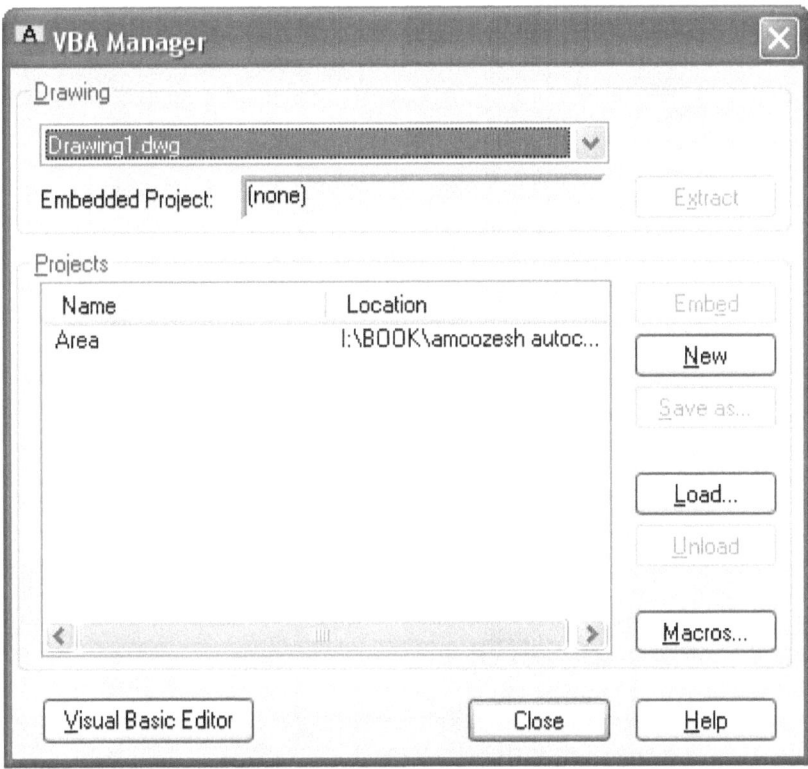

Figure M15

In this window, press the *Load* button and select your file. After seeing the name and path of the file in the *Projects* box, click on the *Macros...* button to open a window named *Macros*.

In this window, press the *Run* button to execute the macro.

The next step is to start creating a button for quick execution of the loaded LISP or Visual Basic program (macro). When creating a button for a LISP file, the code in the macro field should be *^C^C* followed by the shortcut name of the LIPS file.

For example, to create a button for the LISP file named *pickarea* (included in the attached disk), you have to type *^C^Cpickarea* in the macro field of the button (Figure M16).

⊟ **Command**	
Name	**pickarea**
Description	
Extended Help File	
Command Display Name	
Macro	^C^Cpickarea
Tags	
⊟ **Advanced**	
Element ID	MM_190_FF553
⊟ **Images**	
Small image	io.bmp
Large image	

Figure M16

But if the file is made with Visual Basic, the code in the macro field of the new button should be ^C^C*vbarun* followed by a space and then the name of the macro. For example, to create a button for the macro *area* (included in the attached disk), the code in the macro field should be ^C^C*vbarun area* (Figure M17).

⊟ **Command**	
Name	**area**
Description	
Extended Help File	
Command Display Name	
Macro	**^C^Cvbarun area**
Tags	
⊟ **Advanced**	
Element ID	MM_190_18901
⊟ **Images**	
Small image	
Large image	

Figure M17

86-Quick adjustment of the number of decimal places displayed for length and coordinate values

You can use the command Luprec to set the number of decimal places displayed for length, area, and coordinates figures to any value between 0 and 8, without changing the AutoCAD settings or using the *Unit* window. To do this, run the shortcut command *Luprec* in the command line. The software will then ask you to *Enter new value for LUPREC <8>*. In response, you must type the desired number of decimal places, which can be any value between 0 and 8.

Note that this command does not change the actual precision of measurements, but only the number of decimal places displayed.

87-Quick adjustment of the number of decimal placed displayed in angle values

You can use the command *Auprec* to set the number of decimal places displayed for angle values. To do this, execute this command and then type a value between 0 and 8 as the number of decimal places you want to be displayed for angle values.

88-Quick adjustment of the number of decimal places displayed for a specific dimension

Using the Aidimprec command, you can change the number of decimal places displayed for a specific length or angle value. After typing and executing this command, the message *Enter option [0/1/2/3/4/5/6]* will appears in the command line. In response, type the number of decimal places you want to be displayed (could be between 0 and 6) and press enter. Then, click on the dimension you want to edit and press enter again to see the desired changes applied.

89-Changing how a dimension is displayed

You can use the command Aidimtextmove to change how a dimension is displayed in the drawing. This command allows you to select 3 modes of display.

After executing this command, you will be asked to *Enter option [0/1/2]*. In response to this message, you must enter 0, 1, or 2 depending on the display mode that best suits your needs. These modes are illustrated in Figure M18.

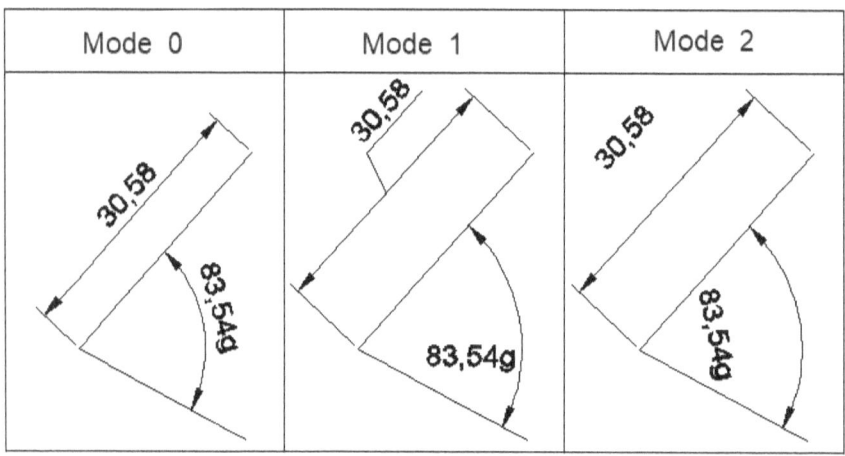

Figure M18

90-Changing the elevation scale of longitudinal and transverse profiles

In road construction projects, it is typical to draw longitudinal profiles with vertical exaggeration ratios of 5, 10, or 20. In cases where surveyors need to change the vertical scale of such longitudinal or transverse profiles, this cannot be done with the *Scale* command, because this command changes the scale in both x and y directions. In order to change the scale along only the y-axis (i.e. to change only the elevation scale of a longitudinal profile), you need to use a command named *Insert Block*. Before learning how to use this command, open and see the files named *Along Section* and *Cross Section* in the folder *dwg*. In these files, longitudinal and transverse profiles are drawn with a vertical exaggeration ratio of 10 (the vertical scale is 10 times the horizontal scale). But as you know, transverse profiles should be drawn on a scale of 1:1. Thus, the vertical (elevation) scale of the transverse profiles must be decreased by a factor of 10.

To do this, open the *Insert* menu and select *Block*, or just execute the shortcut command *I* to open the window shown in Figure M19. Press the *Browse* button and select the file *Cross Section*.

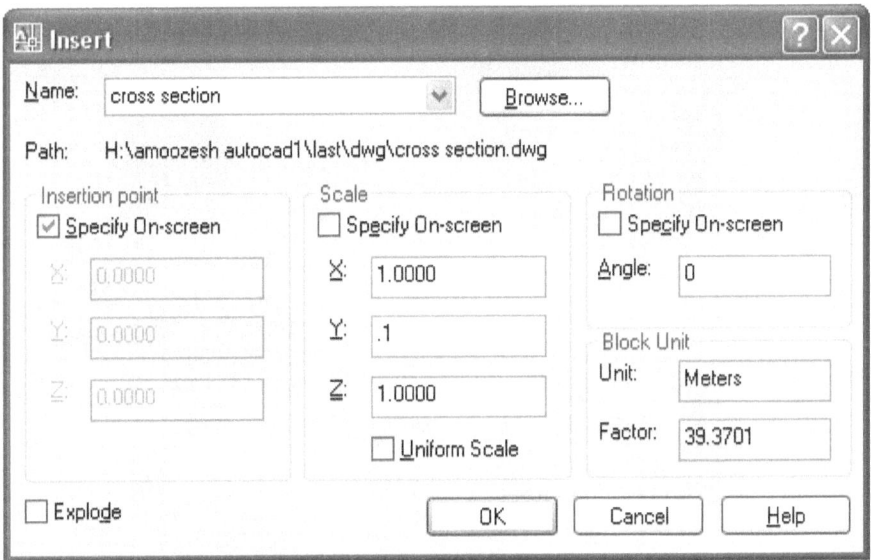

Figure M19

Since we need to reduce the vertical scale by a factor of 10, the *Y* field in the *Scale* box should be set to 0.1. Press OK to close the window, then click on the point where you want the block inserted. By doing this, you will insert all cross sections with a scale of 1:1.

Note that there is actually no need to open the source file and you can do this in the file on which you are currently working. Also, remember that if you adjust the scale settings as explained above, all inserted objects will be rescaled, so to avoid error, it is recommended to first isolate the profiles in another file, and then use that file for insertion.

After insertion, you will see that the inserted texts are also rescaled and thus difficult to read. Hence, the next goal is to edit these texts to restore them to their original state.

To do this, use a selection method such as Quickselect to select the texts, then execute the command *PR* to open the *Properties* window (Figure M20). In the *Text* box of this window, set the *Style* to *Standard*. If this does not fix the problem, return to this window and reduce the *Width factor* to one-tenth of its current value.

Before doing this, however, you need to use the *X* command to explode the inserted blocks (profiles).

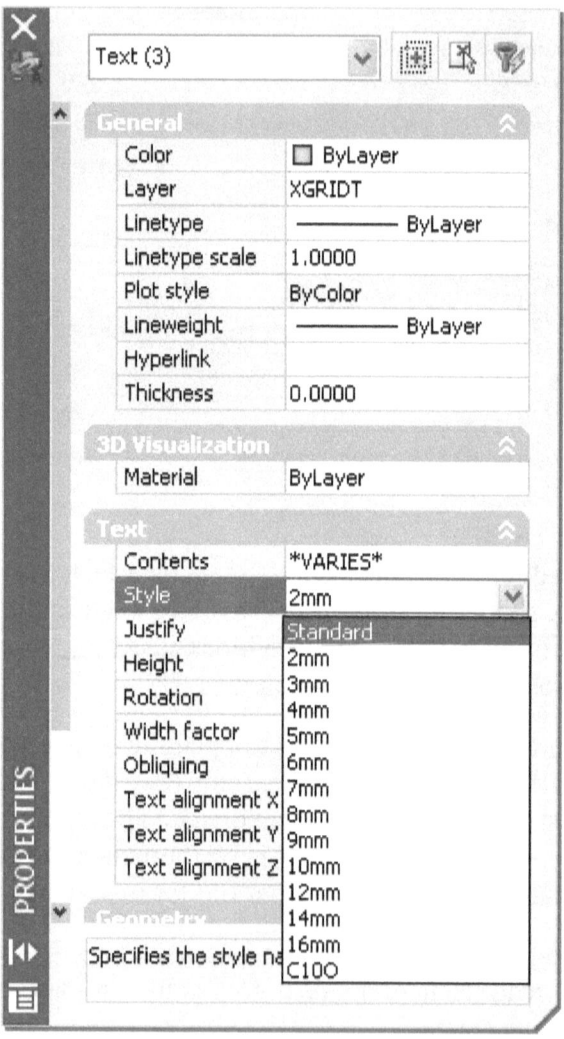

Figure M20

91-Transformation of local coordinates to global coordinates

Definition of transformation:

In the field of surveying, transformation means converting the coordinate system from one mode to another.

A transformation may become necessary for two reasons:

- To convert local coordinates to UTM coordinates

- When the survey is done with a Total Station configured incorrectly for any reason.

Before explaining the transformation process, it is worth to quickly review the stage of surveying. These stages are as follows:

• Reconnaissance and marking

• Traversing and error balancing

• Levelling

• Feature surveying

• Drawing and cartography

In the traversing stage, stations can be converted to UTM coordinates based on the horizontal coordinates provided by a national mapping agency. This can be done in both land and satellite surveying.

In land surveys, coordinates are often measured with angle and distance measurement instruments. But in satellite surveys, these measurements are carried out using single-frequency and dual-frequency GPS. To accelerate the surveying operations, the steps after reconnaissance and marking, i.e. traversing, levelling, and feature surveying can be performed simultaneously by three separate teams. But how can we start surveying features before finishing traversing and error balancing, i.e. without having the exact coordinates of stations?

When the UTM and exact coordinates of two or more stations are not yet available, we can proceed to survey with local coordinates and transform the collected data to UTM once the exact and corrected coordinates of all stations are available. The standard method of transformation is to conduct separate calculations based on the raw survey data (measured distances and horizontal and vertical angles) of each individual stations with the help of standard computing software such as SDR Map and Excel. But in small projects where there is no need to produce text files for original and transferred coordinates, the whole operation can be performed faster and more conveniently in AutoCAD, where there is no need to repeat the tasks for each individual station. In computational transformation, we often use the coordinates of two points, namely the current station and the backsight point, but in AutoCAD, we use three points to ensure greater precision. It should be explained that the transformation operation in AutoCAD can be described as the combined use of three commands: *Move*, *Rotate*, and *Scale*. In the following, we explain how to perform this transformation with the help of an example.

Open the file Transfer2.dwg in the attached disk. In this example, an area has been surveyed with a local coordinate system consisting of the stations Z1, Z2, Z3, and Z4. After traversing and coordinate transfer, the UTM coordinates of the stations have been obtained as listed in Table M21.

Point	LOCAL		UTM	
	X	Y	X	Y
Z1	1768.9217	6490.9361	625705.7785	3077695.9900
Z2	2070.1435	5989.0668	625676.0257	3077111.5399
Z3	2861.8440	6574.9870	626658.0534	3077153.1677
Z4	2540.4905	7019.1818	626640.8143	3077701.1319

Table M21

Here, we use the points Z1, Z2, and Z3 for transformation and use Z4 for control. For control, the transferred coordinate of Z4 after transformation will be compared with the UTM coordinates of the same point. For transformation in AutoCAD, we use the command *Align* with the shortcut *AL*. But before doing this, we have to draw the stations with the *Point* command, so that they can be controlled after the transformation. To do so, type and execute the shortcut command *PO* and then type the UTM coordinates of Z1-Z4 to draw them. Then, select all the objects on the drawing including the station points and execute the *AL* command in the command line. The software will then ask you to *Specify first source point*. In response, enter the local coordinates of Z1. Then, the software will then display the message *Specify first destination point*, in response to which you must enter the UTM coordinates of Z1. The next two massages will ask you to *Specify second source point* and *Specify second destination point*, in response to which you must enter the local coordinates and UTM coordinates of Z2. After specifying the second point, the massage *Specify third source point or <continue>* appears. In this step, you can decide whether to specify a third pair of source and destination points or continue the process with just two points. For this example, use the local and UTM coordinates of Z3 as the third source point and the third destination point respectively. In short, the contents typed in the command line should be as follows.

Command : al

Command : Specify first source point: 1768.9217,6490.9361

Command : Specify first destination point: 625705.7785,3077695.99

Command : Specify second source point:2070.1435,5989.0668

Command : Specify second destination point:625676.0257,3077111.5399

Command : Specify third source point or <continue>:2861.844,6574.9870

Command : Specify third destination point:626658.0534,3077153.1677

After specifying the third point, the points will be automatically transformed. If you do not see the transformed plot, use the command *Zoom All*. The next step is to use the point Z4 to control the accuracy of work. If you completely zoom on Z4, you will notice that the transferred Z4 is only slightly different (about 2 centimeters) from its UTM counterpart, which is sufficiently accurate for most projects.

To practice, repeat the transformation with the points Z2, Z3, and Z4 and use Z1 for control.

92-Drawing traverses

First, it is important to note that AutoCAD does not have a specific command for drawing traverses, and traverse calculation and drawing are typically performed with other software such as Excel, SDR map, Survey, etc. However, you can AutoCAD commands to draw simple traverses. Also, remember that AutoCAD is not recommended for error balancing, so this exercise assumes that the error of closure is acceptably small. Rather than being focused on traverses, this exercise aims to familiarize the reader with a number of tricks and shortcuts that make it easier to work with AutoCAD.

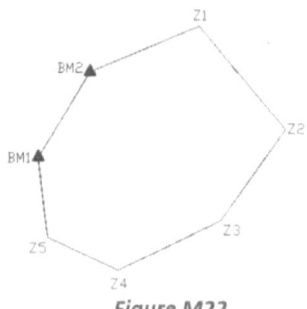

Figure M22

Coordinates of benchmark points			Distances		Angles	
ID	X	Y	ID	Value	ID	Value
BM1	565099.3251	4157280.6352	BM2-Z1	768.6523	BM2	160.3785g
			Z1-Z2	861.3194	Z1	120.6906g
			Z2-Z3	714.0180	Z2	115.1927g
BM2	565442.4798	4157822.6317	Z3-Z4	742.7243	Z3	167.2654g
			Z4-Z5	512.3764	Z4	145.5294g
			Z5-BM1	508..9949	Z5	134.2081g
					BM1	156.7353g

Table M23

In this exercise, we draw a traverse with the general layout displayed in Figure M22. In this traverse, the benchmark points BM1 and BM2 have been used to measure the coordinates of unknown points Z1-Z5. All angle and distance values have been recorded with 0.0001 precision. Each angle has been measured with four repeats of double-centering and each distance has been measured eight times. The angle and length values obtained after error balancing are given in Table M23.

First, to plot the baseline BM1-BM2, the software coordinate system must be set to the absolute Cartesian mode (because BM1 and BM2 both have UTM coordinates, which is an absolute Cartesian system). To do this, run the shortcut command *SE* to open the *Drafting Settings* window. Go to the *Dynamic Input* tab and click on the *Setting* button in the *Pointer Input* pane to open the *Pointer Input* window. In this window, tick the *Cartesian Format* and *Absolute Coordinate* checkboxes, and press OK to confirm the selections. Next, you need to draw the benchmark points with the *Line* command. For this purpose, run the shortcut command *L* and then enter the coordinates of BM1 as the first point, enter the coordinates of BM2 as the second point, and finally press enter to draw a line between these points. If you do not see the line, run the command *Zoom Extend*.

Command: L

Command: LINE Specify first point: 565099.3251 , 4157280.6352

Command: Specify next point or [Undo]: 565442.4798 , 4157822.6317

Command:Z

Command:E

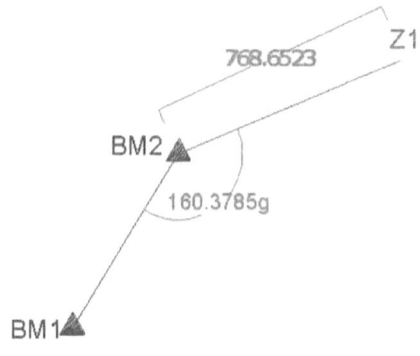

Figure M24

After drawing the benchmark points (BM1, BM2), you must use the collected distance and angle data to draw traverse vertices one by one. Before doing this, you have to set the coordinate system to the relative polar coordinate system. To do so, run the *SE* command, go to the *Dynamic Input* tab and enable the *Polar Format* and *Relative Coordinate* options in the *Pointer Input* pane.

Now you must use the BM2-Z1 distance and the BM2 angle to find the position of Z1 relative to BM2. Note that since Z1 should be drawn based on BM2, you have to set the reference of angle measurement to BM2-BM1 line. To do this, execute the shortcut command *UN* in the command line to open the *Drawing units* window. In this window, set the unit of angle measurement to grad with 4 decimal places (because in this exercise, angles have been measured in grads with 4 decimal places). Then, click on the *Direction* button on the bottom of this window to open the *Direction Control* window. In this window, you can set the reference direction of angles to north, east, south, west, or any direction of your choosing. Since we need to set the reference of angle measurement to a specific direction, you have to enable the *Other* option and click on the bottom below it. After pressing this button, the windows will be closed and AutoCAD will be ready to receive the direction to be used as angle measurement reference. In this step, you just have to click first on BM2, and then on BM1 to specify the BM2-BM1 line as the reference of angle measurements. When done, press OK until all windows are closed. Note that if angles were external, you had to choose the BM1-BM2 direction instead of BM2-BM1.

Now, to draw the point Z1, execute the shortcut command *L* and click on BM2 to mark it as the first point. Then, use the sign @ to inform the software that the values to be entered are relative to the first point. Next, type the measured distance between BM2 and Z1, i.e. 768.6623. Then, type the sign < to inform the software that you intend to enter the angle, and then type the BM2 angle, i.e. 160.3785. Press enter when done. Now, you must see a line drawn between BM2 and Z1.

Command: PL

Command: LINE Specify first point: (click on BM2)

Command: Specify next point or [Undo]: @768.6623<160.3785

After drawing the BM2-Z1 line, you must repeat the above process to draw the Z1-Z2 line using the measured distance between Z1 and Z2 and the measured Z1 angle (remember to change the angle reference to the Z1-Z2 line). Repeat this process until reaching the point BM1 and forming a closed traverse. Open traverses can also be drawn in the same way.

93-Quick insertion of feature symbols using blocks

One of the basic requirements for standard drawings is to use standard symbols for features. The standard symbols defined for features such as light poles, trees, wells, etc., are provided in the final chapter of the book.

To insert the symbol of a feature, you can simply copy the symbol and paste it at the desired points. But this cannot be done in larger drawings, where the number of features is quite high, because manual insertion of feature symbols will be extremely time-consuming.

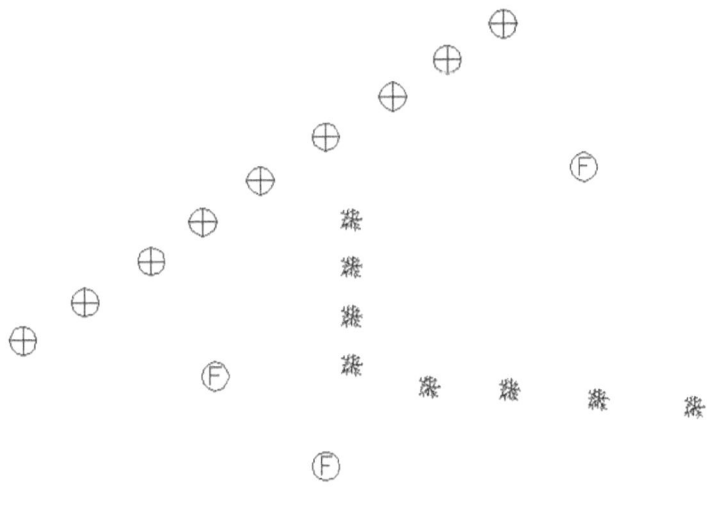

Figure M25

Specialized survey software such as *SDR map* and *Land* have some features for automatic insertion of these symbols. However, in cases where the symbols included in the software are not suitable for our project or the software do not allow new symbols to be defined, we have to use AutoCAD to insert feature symbols.

In AutoCAD, you can use blocks for this purpose. To do this, first, you must use the other software to insert a block for each type of feature, then change these blocks to the desired symbol in AutoCAD.

Consider the drawing shown in Figure M25. Produced by *SDR map*, this drawing contains 9 light poles, 3 wells, and 8 trees. In *SDR map*, each of these features has been given a Block-type symbol with the following names.

Feature Name	Symbol	Block Name
Tree	❄	**TREE_2-12**
Well	Ⓕ	**SS_MANHOLE-20**
Light pole	⊕	**CROSS_CICLE-18**

Open the file Symbole.dxf in the folder dwg in the attached disk.

This file contains the standard symbols of the above feature symbols (Figure M26).

Figure M26

These symbols are for well, tree, and light pole respectively. Now, the goal is to use these symbols in place of previous symbols.

94-Mass importing of points into AutoCAD with the help of Excel

In this part of the book, we explain how you can use Excel to accelerate your work in AutoCAD.

It should be noted that Microsoft Excel is a very powerful and practical software with wide-ranging applications in survey computations (e.g. for traverse calculations, error balancing in levelling projects, etc.), which fall beyond the scope of this book. With that said, this section is focused on the simple solution that can be implemented in Excel to speed up the process of executing frequently used AutoCAD commands.

For example, suppose you want to create a single point in AutoCAD. To do this, you must first type the command *PO* or *Point* in the command line, press Space or Enter to execute it, then type X, Y, and Z coordinates of the point separated by a comma, and finally press Space or Enter to draw the point.

Now, suppose you want to draw two points with the following coordinates.

| 1 | 675000 | 3824200 | 1311.51 |
| 2 | 674960 | 3824200 | 1311.43 |

If you manage to produce a text or Excel file with the contents:

Point 675000,3824200,1311.508

point 674960,3824200,1311.428

You can simply copy this into the command line and create both points at once (because the text in each line satisfies the requirements for running the *Point* command).

Practice this on the file named 2point.txt provided in the data folder.

Important note

In the AutoCAD command line, *Space* key performs the same function as *Enter* key. In other words, this key executes a typed command and if no command is typed in, it executes the last command executed. Hence, we can use the *Space* character to emulate the act of pressing *Enter* after typing a command. In the above lines of text, one space character is used after the word *Point* to start executing this command and another is used after the Z coordinate to finish the command. Thus, both of them are necessary for executing the *Point* command.

The explained method can be used to draw a large number of points at once, but preparing the text needed to do this could become very time-consuming. Fortunately, we can use Excel to speed up the process and create a large body of text ready for use in the AutoCAD command line.

Note that to use Excel, you must first learn the rules that apply to writing texts and defining functions in this software. These rules can be learned, rather easily, from Excel training books available online or in paper. For example, to produce the above lines of text, you must create a function in the following form:

="point "&x&","&y&","&z

where x, y, and z are the cells containing the X, Y, and Z coordinate of the point.

Suppose you want to import the points of Figure M27 using the above formula.

	A	B	C	D	E
					="point "&B1&","&C1&","&D1
1	1	675000	3825320	1312.01	="point "&B1&","&C1&","&D1
2	2	674960	3825320	1312.025	
3	3	674760	3825280	1313.96	
4	4	674800	3825280	1314.615	
5	5	674840	3825280	1315.59	
6	6	674880	3825280	1316.565	
7	7	674920	3825280	1317.54	
8	8	674960	3825280	1312.01	
9	9	675000	3825280	1312.025	
10	10	675000	3825240	1313.96	
11	11	674960	3825240	1314.615	
12	12	674920	3825240	1315.59	
13					

Figure M27

To do this, after inserting the points into Excel, type the above formula in the first row of a new column (in this example, the cell E1) and then replace the x, y, and z with the name of the cells containing the X, Y, and Z coordinates, which here are B1 and C1 and D1. After editing the formula, press enter to confirm it. The contents of the cell E1 should then change to:

point 675000,3825320,1312.01

Now, move the mouse cursor to the bottom right corner of the cell E1 until the cursor changes to a + sign, and then click and drag down until reaching the last row of data, that is, the cell E12. This makes the software reproduce the formula for all the subsequent rows, as shown in Figure M28.

	A	B	C	D	E
1	1	675000	3825320	1312.01	point 675000,3825320,1312.01
2	2	674960	3825320	1312.025	point 674960,3825320,1312.025
3	3	674760	3825280	1313.96	point 674760,3825280,1313.96
4	4	674800	3825280	1314.615	point 674800,3825280,1314.615
5	5	674840	3825280	1315.59	point 674840,3825280,1315.59
6	6	674880	3825280	1316.565	point 674880,3825280,1316.565
7	7	674920	3825280	1317.54	point 674920,3825280,1317.54
8	8	674960	3825280	1312.01	point 674960,3825280,1312.01
9	9	675000	3825280	1312.025	point 675000,3825280,1312.025
10	10	675000	3825240	1313.96	point 675000,3825240,1313.96
11	11	674960	3825240	1314.615	point 674960,3825240,1314.615
12	12	674920	3825240	1315.59	point 674920,3825240,1315.59
13					

Figure M28

Now, to insert your points, you just have to select and copy these cells (Figure M29) and then paste them in the AutoCAD command line (by right-clicking on the command line and selecting *Paste* or clicking in the command line and pressing Ctrl+V)

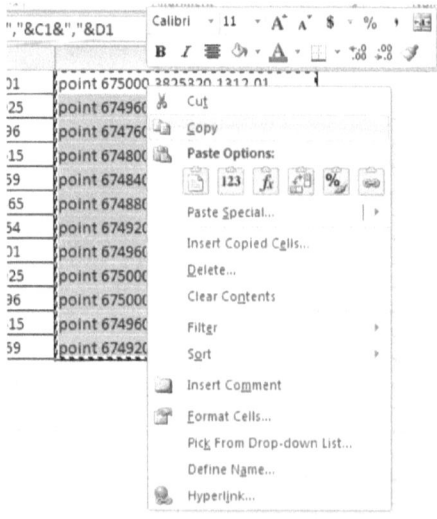

Figure M29

You can use the file point.xls in the folder data to practice this method.

To do this, first, select all objects of the light pole, and then execute the shortcut command *B* (for *Block*). Next, press the *Pick point* button in the *Base point* box and click on the block's base point to specify it as the base point for insertion. Now, to replace the current symbol with the new one, you must type the name of the current symbol in the *Name* field (here CROSS_CIRCLE-18) and press OK.

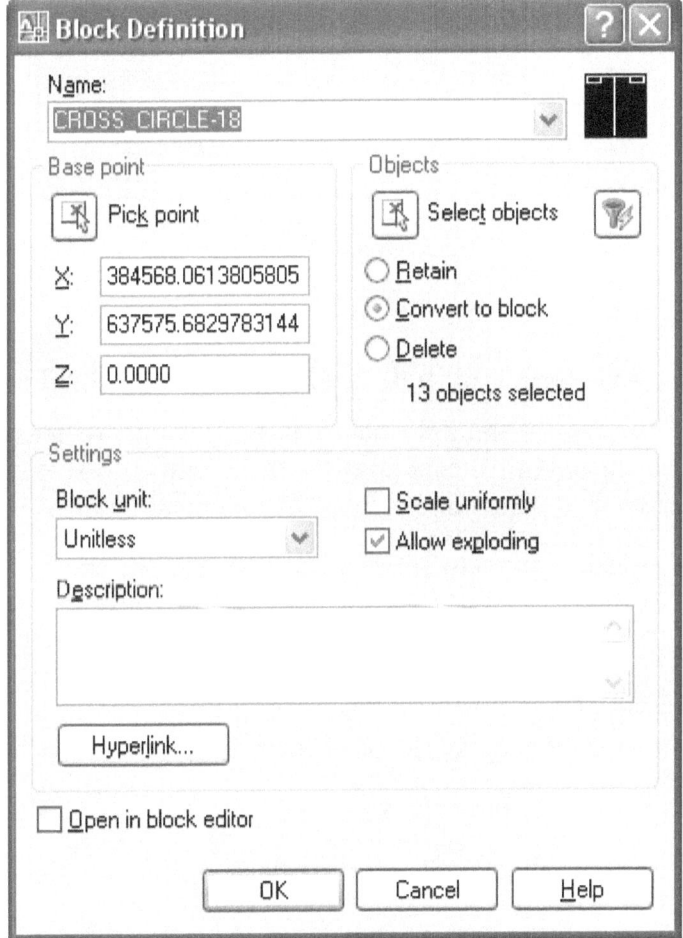

Figure M29-30

After closing this window, you will be warned that the block with the specified name is already defined and will be asked whether you want to update the existing blocks (Figure M30). If you Press Yes, the software will replace the current blocks with the new blocks, which is exactly what we wanted to do from the start. Finally, you just have to repeat this process for other symbols.

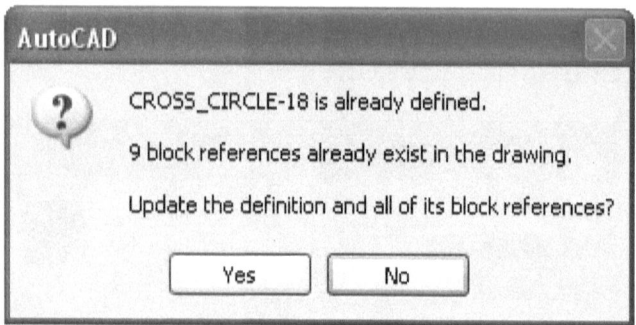

Figure M30

95-Creating texts for a multitude of points with the help of Excel

To create a text for multiple points in a drawing, we must examine how texts are created in AutoCAD, emulate the steps in Excel, and finally copy the results from Excel into AutoCAD command line.

Steps that must be taken to create a text in AutoCAD are:

1. Type the command -text

2. Type a space to execute the command

3. Enter the x coordinate of the base point

4. Type a comma character ((,))

5. Enter the y coordinate of the base point

6. Type a space to confirm the base point and proceed to the next step

7. Enter the text height

8. Type a space

9. Enter the text rotation angle (usually zero)

10. Type a space

11. Enter the desired text (can be the point elevation, code, coordinates, or any other data)

The excel formula for emulating the above steps is:

="-text "&X&","&Y&" "&height&" "&angle&" "&text

where X and Y are the cells containing the x and y coordinates of the point, *height* is the desired text height, and *angle* is the desired text rotation angle.

For example, to perform this process for the points of the previous exercise, you should create an excel file like the one displayed in Figure M31.

	A	B	C	D	E
				CONCATENATE ▾ × ✓ ƒ	="-text "&B1&","&C1&" "&2&" "&0&" "&D1
1	1	675000	3825320	1312.01	="-text "&B1&","&C1&" "&2&" "&0&" "&D1
2	2	674960	3825320	1312.025	-text 674960,3825320 2 0 1312.025
3	3	674760	3825280	1313.96	-text 674760,3825280 2 0 1313.96
4	4	674800	3825280	1314.615	-text 674800,3825280 2 0 1314.615
5	5	674840	3825280	1315.59	-text 674840,3825280 2 0 1315.59
6	6	674880	3825280	1316.565	-text 674880,3825280 2 0 1316.565
7	7	674920	3825280	1317.54	-text 674920,3825280 2 0 1317.54
8	8	674960	3825280	1312.01	-text 674960,3825280 2 0 1312.01
9	9	675000	3825280	1312.025	-text 675000,3825280 2 0 1312.025
10	10	675000	3825240	1313.96	-text 675000,3825240 2 0 1313.96
11	11	674960	3825240	1314.615	-text 674960,3825240 2 0 1314.615
12	12	674920	3825240	1315.59	-text 674920,3825240 2 0 1315.59
13					

Figure M31

Next, you just have to select and copy the created cells and paste them in the AutoCAD command line.

96-Fast drawing of a route's axis line with the help of Excel

In many projects, you will need to draw the axis line (centerline) of a route with *Line* or *Pline* commands. In cases where the points are in correct order, such lines can be drawn very quickly with the help of Excel.

Suppose that the points with the coordinates given below constitute the axis line of a route.

1	311.072	962.631
2	361.058	961.521
3	411.057	961.43
4	461.013	963.357
5	510.74	968.439
6	559.797	977.978
7	607.333	993.367
9	651.999	1015.704
10	692.16	1045.37
11	725.865	1082.195
12	751.058	1125.263
13	766.633	1172.697
14	776.544	1221.694
15	787.185	1270.537
16	807.576	1315.878
17	849.101	1341.599
18	898.826	1345.983
19	948.79	1347.548

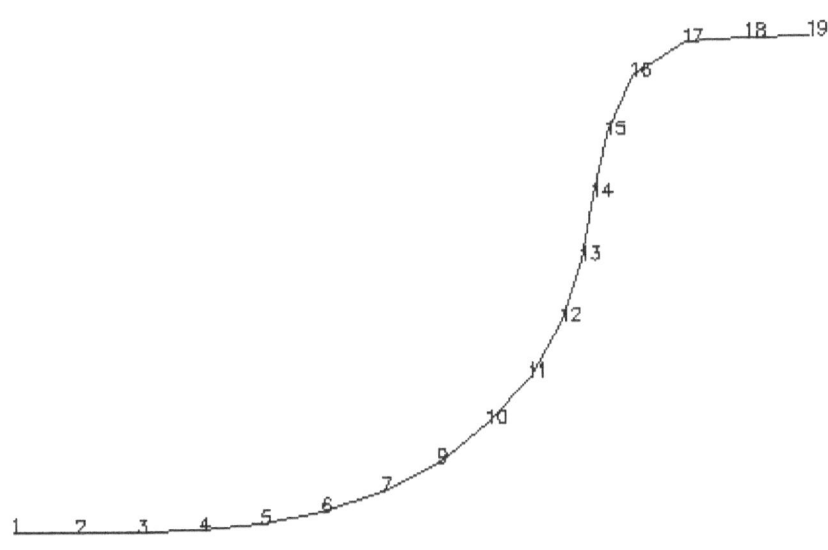

Figure M32

As can be seen in Figure M32, these points are in the correct order and are numbered 1 to 19, respectively. The Excel formula for drawing a line with these points using the *Pline* command is:

="pline "&X&","&Y

where *X* and *Y* are the cells containing the x and y coordinates of the point. The rest of the process is similar to what was explained in previous examples. As before, you must select and copy the resulting cells and paste them in the AutoCAD command line.

For more practice, use the file Ax.xls in the folder data.

	A	B	C	D
		CONCATENATE	▼ ⊙ ✕ ✓ *fx*	="pline "&B1&","&C:
	A	B	C	D
1	1	311.072	962.631	="pline "&B1&","&C1
2	2	361.058	961.521	pline 361.058,961.521
3	3	411.057	961.43	pline 411.057,961.43
4	4	461.013	963.357	pline 461.013,963.357
5	5	510.74	968.439	pline 510.74,968.439
6	6	559.797	977.978	pline 559.797,977.978
7	7	607.333	993.367	pline 607.333,993.367
8	9	651.999	1015.704	pline 651.999,1015.704
9	10	692.16	1045.37	pline 692.16,1045.37
10	11	725.865	1082.195	pline 725.865,1082.195
11	12	751.058	1125.263	pline 751.058,1125.263
12	13	766.633	1172.697	pline 766.633,1172.697
13	14	776.544	1221.694	pline 776.544,1221.694
14	15	787.185	1270.537	pline 787.185,1270.537
15	16	807.576	1315.878	pline 807.576,1315.878
16	17	849.101	1341.599	pline 849.101,1341.599
17	18	898.826	1345.983	pline 898.826,1345.983
18	19	948.79	1347.548	pline 948.79,1347.548
19				
20				

Figure M33

97-Quick insertion of feature symbols with the help of Excel

Consider the following survey table.

1	384027.71	637561.75	TIR
2	384045.08	637571.98	TIR
3	384063.69	637583.14	TIR

4	384078.25	637593.67	TIR
5	384094.39	637604.83	TIR
6	384112.69	637616.61	TIR
7	384131.63	637627.45	TIR
8	384146.81	637637.37	TIR
9	384162.32	637646.97	TIR
10	384081.35	637552.08	CHAH
11	384184.75	637608.10	CHAH
12	384112.21	637527.51	CHAH

This table contains the coordinates of the points in Exercise 7, which are related to three types of feature, namely light poles (TIR), water wells (CHAH) and trees (DERAKHT). Here, the goal is to insert the symbols of each feature on the related points.

To do this, we will use the *Copy* command in AutoCAD and then using the destination coordinates prepared in an Excel file. In AutoCAD, after copying an object and pressing enter, the software will ask you to spccify the point where it must bc pasted, either by mouse clicking or by entering the coordinates in the command line. Hence, you can copy a symbol with the *Copy* command, then copy/paste a list of points where symbol must be pasted from an Excel file. To prepare this list in excel, you can use the following simple formula:

=X&","&Y

Note that this process should be done separately for each feature. Thus, you must first prepare a list of points with the code representing the feature, select and copy the corresponding symbol, and paste it at the coordinates of the prepared list. To practice, use the file block.xls in the folder data, which contains the above points, and the file Symbol.dwg in the folder dwg, which contains the symbols of the features located at these points.

98-Using Excel's Concatenate function to construct AutoCAD commands

You can use the Concatenate function in excel to carry out the operations of exercises 8 to 11 even more quickly. Suppose we want to use with this option to complete Exercise 8, i.e., import multiple points into AutoCAD. To do this, you must open the excel file, click on the first cell of a new column, and then click on the *f*$_x$ button or press Shift+F3. In this window (Figure M34), set the *Or select a category* menu to *All* and select the *CONCATENATE* function from the function list. Press OK to open the *Function Arguments* window displayed in Figure M35.

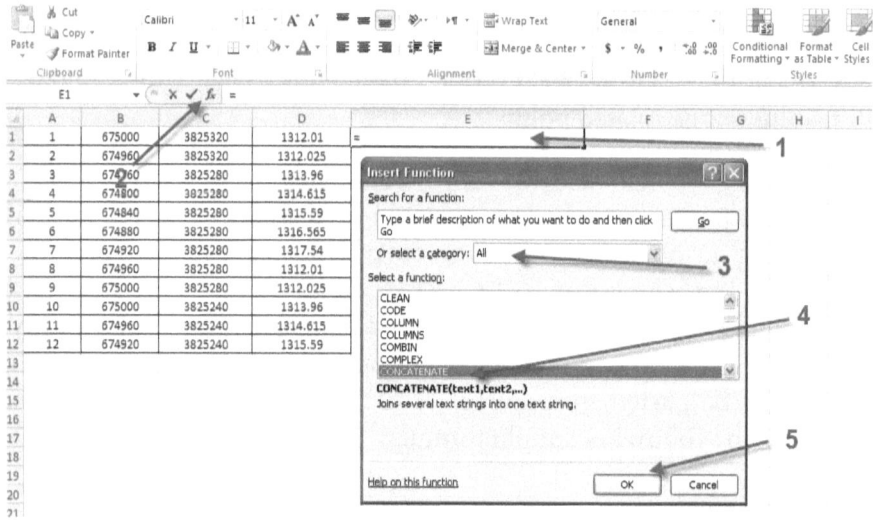

Figure M34

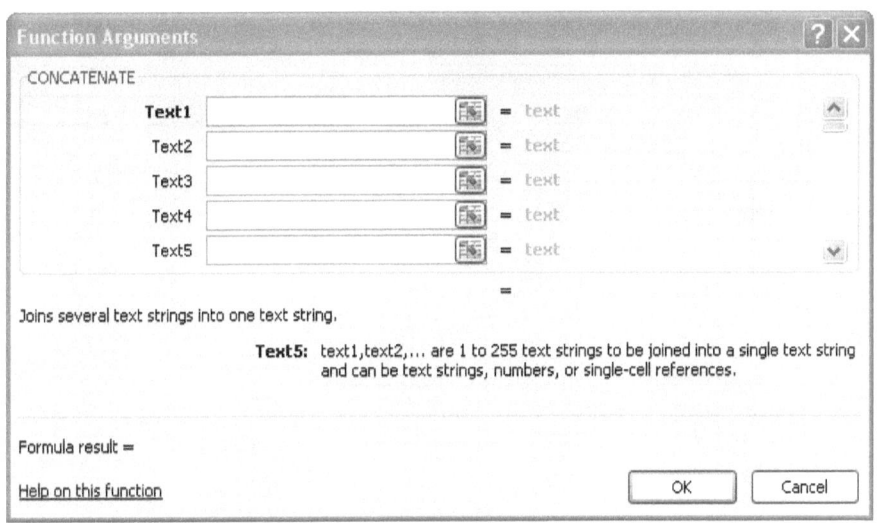

Figure M35

This window contains up to 255 boxes, named *Text1*, *Text2* .... , which can be used to construct a formula for creating AutoCAD commands. For this purpose, you must enter in the text boxes the words and components that will constitute the desired AutoCAD command. In the case of Exercise 8 for example, you must type *point* in *Text1* box, type a *space* character in *Text2* box, press the *Text3* button and select the cell holding the first x-coordinate, type a *comma* character in *Text4* box, press the *Text5* button and select the cell holding the first y-coordinate, type another *comma* character in *Text6* box, press the *Text7* button and select the cell holding the first z-coordinate, and type another *space* character in *Text8* box. When done, press OK to create the *Concatenate* function, as shown in Figure M36.

	Clipboard	⌐		Font		⌐	Alignment

	E1	▾ ⊂	ⁿ	fx	=CONCATENATE("point"," ",B1,",",C1,",",D1," ")		

	A	B	C	D	E
1	1	675000	3825320	1312.01	point 675000,3825320,1312.01
2	2	674960	3825320	1312.025	
3	3	674760	3825280	1313.96	
4	4	674800	3825280	1314.615	
5	5	674840	3825280	1315.59	
6	6	674880	3825280	1316.565	
7	7	674920	3825280	1317.54	
8	8	674960	3825280	1312.01	
9	9	675000	3825280	1312.025	
10	10	675000	3825240	1313.96	
11	11	674960	3825240	1314.615	
12	12	674920	3825240	1315.59	

Figure M36

You can now drag down the created cell to let the software reproduce the function for the subsequent rows. When finished, copy the resulting column and paste it into the AutoCAD command line. To practice, redo the exercises 8-11 using the *Concatenate* function.

99-Digitization of longitudinal profiles

By digitizing the longitudinal profile, we mean creating a longitudinal profile where the chainage (station) and elevation data of any point can be easily extracted with the ID command.

For example, Figure XX shows the longitudinal profile of a 360 meters long route, which involves both cut and fill operations and contains two vertical curves.

The important parameters for digitalization of a profile are horizontal and vertical scales, chainage (station), and elevation.

Horizontal and vertical scales are important because a digital profile should have a 1:1 scale in both directions. In the profile of this example, the horizontal scale is 2:1 and the vertical scale is 1:5 and, hence, the horizontal scale should be doubled and the vertical scale should be divided by 5 (or multiplied by 0.2).

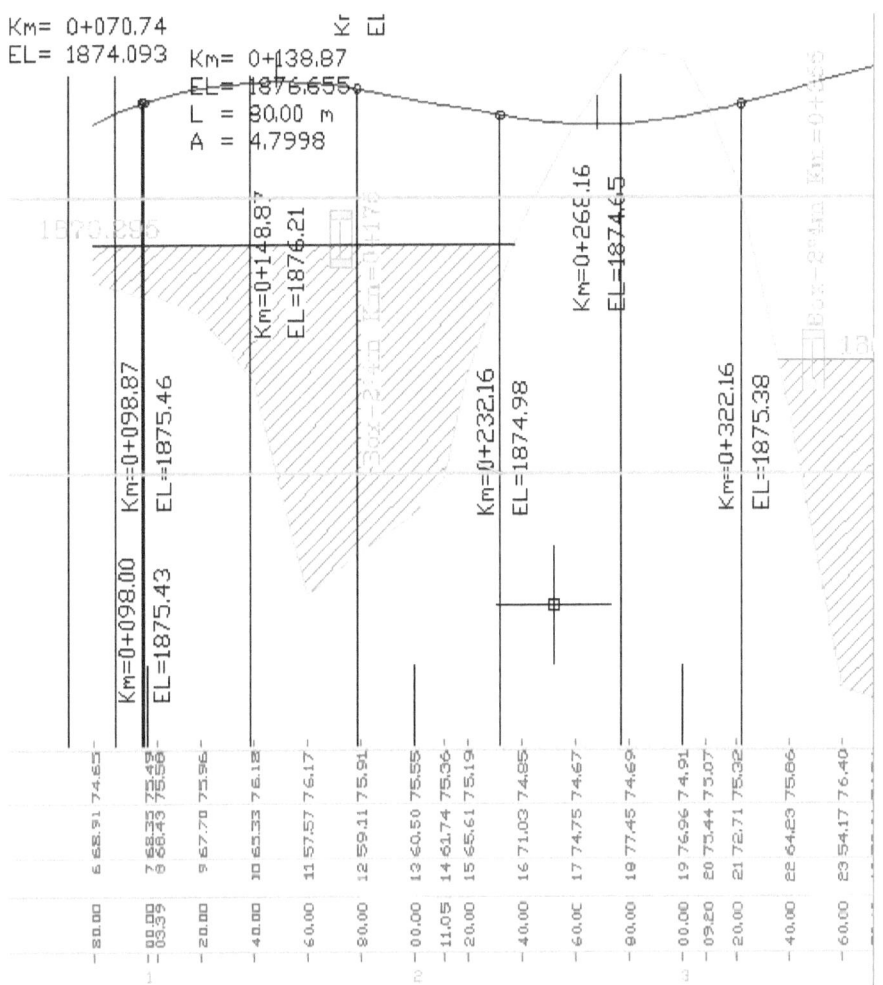

Figure M37

The point of all of this is to relocate the entire profile such that the X and Y coordinates of the points become equal to the route's chainage and elevation values. For example, the station 0+200 of the profile displayed in Figure M37 has an elevation of 75.55. Thus, if we transfer the profile such that this point fall on the coordinates (200, 75.55), then we have created a digital profile where station and elevation of any point can be easily determined by checking its X and Y values.

This can be done faster and more conveniently with the help of blocks. For this purpose, execute the shortcut command B in the command line to open the *Block Definition* window (Figure M38).

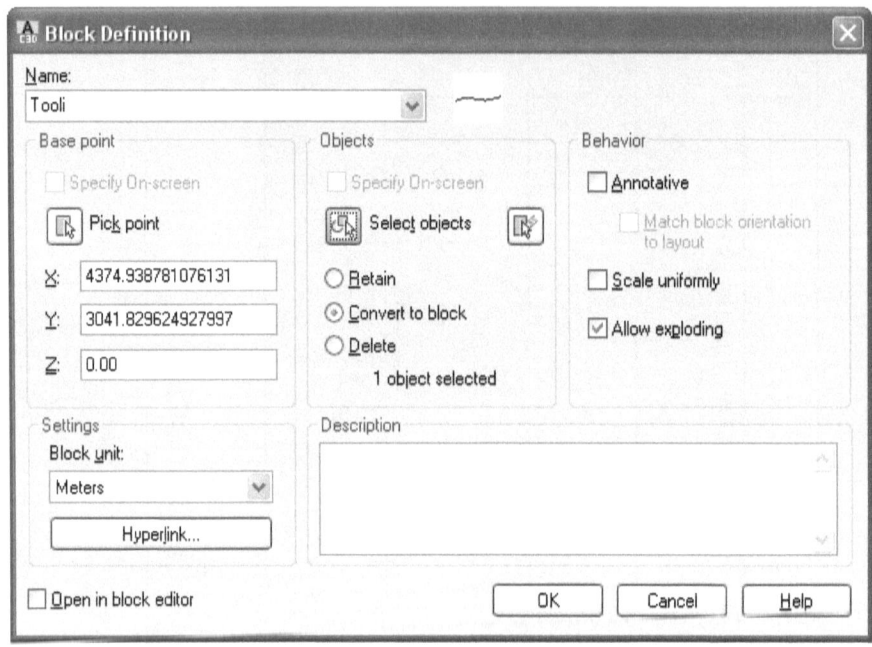

Figure M38

In this window, first type a name for the block, then use the *Select objects* button to select the project line (the blue line). After selecting all project lines, press the *Pick point* button, click on the 0+200 station on to the project line. In this step, make absolutely sure to click exactly on the point on the route, because the entire profile will be transferred based on this point. For an easier selection of this point, you can mark it in advance with a circle or with a colored point.

When done, press OK to make the profile into a block with the base point at (200, 75.55). Note that this point will now have a different coordinate.

The next step is to insert this block on the target point and change its scale. For this purpose, you must use the command *Insert Block* with the shortcut *I*. After executing this command, you will be directed to the *Insert* window shown in Figure M39.

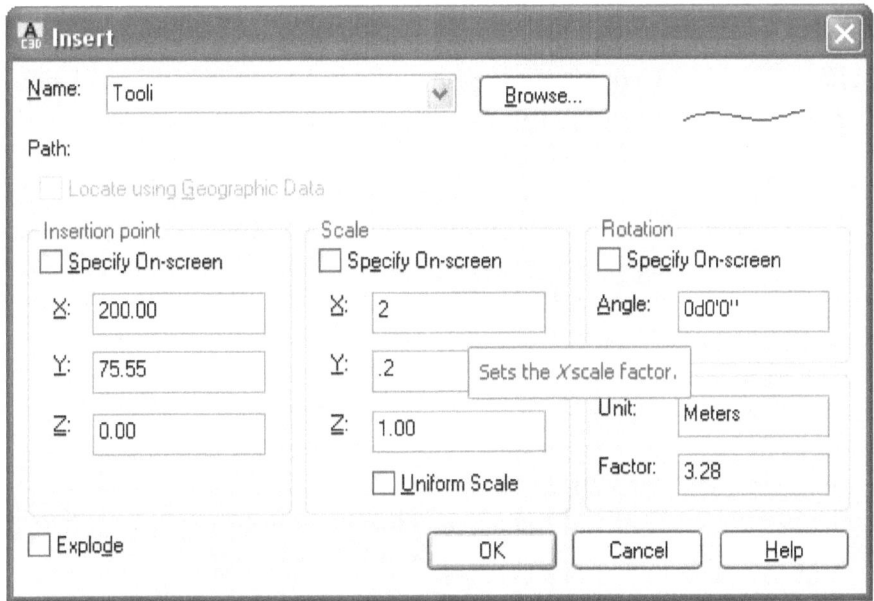

Figure M39

In this window, select the block you just created. In the pane *Insertion point*, you must enter the coordinates (200, 75.55). In the *Scale* pane, the horizontal scale should be doubled and the vertical scale should be multiplied by 0.2.

When finished, press OK to insert the profile at the target point.

Now, you can run the ID command on any point of the profile to determine its station and elevation values.

The files related to this exercise are located in the folder named Long section.

Note that it is customary to use the first point of the profile as the insertion point, but to avoid confusion, here we used an arbitrarily selected point (200, 75.55).

To better understand the process, try to transfer the profile using the coordinates of the first point (80, 74.65).

100-Extraction of project data from longitudinal profiles

In many longitudinal profiles, like the one in the previous exercise, project data are displayed at regular intervals (usually 20m intervals). But to obtain project data from any other station, you have to calculate them for the target point. In straight segments, this calculation is a simple interpolation between the stations positioned immediately before and after the target point. In the longitudinal profile of the previous exercise, for example, the project elevation at the stations 0+160 and 0+180 is 76.17 and 75.91 respectively. Having these values, the project elevation at the station 0+170 can be easily interpolated to 76.04. However, if the road segment is not absolutely straight, we cannot use interpolation in this way and have to conduct curve calculations. Moreover, in cases where the route is long and contains multiple vertical curves, it would be very difficult, time-consuming, and error-prone to manually calculate project elevations at small intervals (sometimes the interval has to be 1 meter or even half a meter, particularly in tunneling projects). AutoCAD does not contain an explicit feature for dealing with this issue, but the problem can be solved by using a combination of AutoCAD commands.

But first, you need to digitize the longitudinal profile as explained in the previous exercise.

After digitizing the profile, use the *Line* command to draw a line below the profile. Then draw another line vertically such that its extension would cross the first point of the profile (see Figure M40).

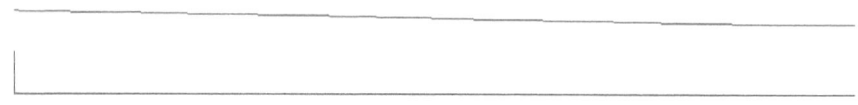

Figure M40

Make sure to draw the vertical line drawn with the *Line* command and none other. It is also recommended to position the line in such a way as to obtain round stations. For example, if your profile starts at a non-round chainage like 0+57.5 and you want the stations to be round, then you must relocate the line by 2.5 meters to make it match the 0+60.0 station.

The next step is to copy this line at the desired station or intervals. Here, we use the *Array* command to accelerate this operation. Executing the *array* command opens the *Array* window displayed in Figure M41 (If you are using 2009 or later versions of AutoCAD, execute the *Arrayclasic* command).

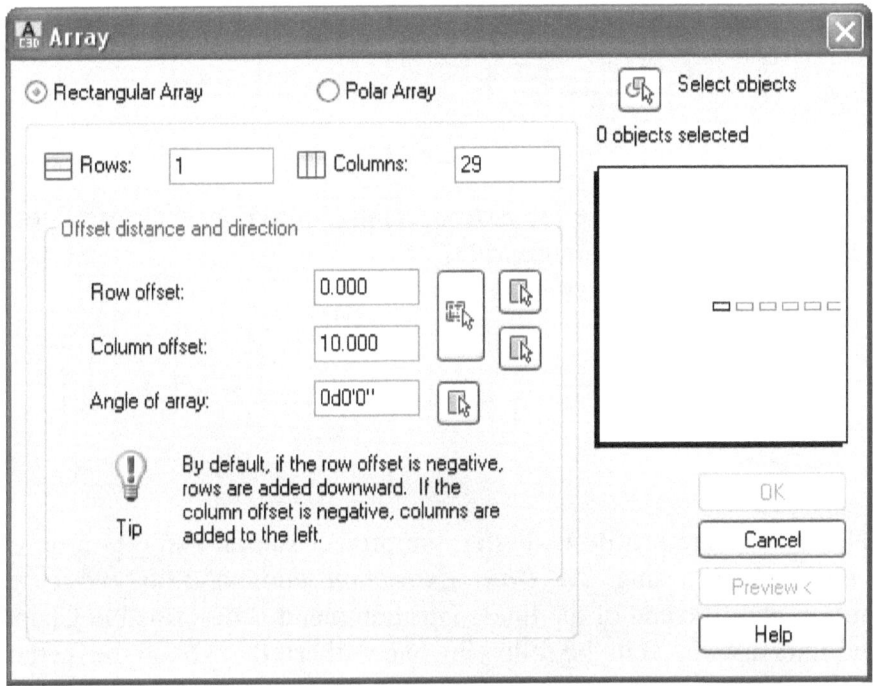

Figure M41

Press the *Select object* button in the upper right part of this window, click on the vertical line you drew earlier, and press enter to reopen this window. Then, use the *Rows* and *Columns* fields to set the number of rows to 1 the number of columns to 29.

The choice of the number of columns depends on the total profile length and the intervals at which we need project data. For example, since this profile is 290 meters long and we want data at 10-meter intervals, the number of columns is set to 290/10=29. Similarly, to have data at 5 meters and 1-meter intervals, this number should be set to respectively 290/5=58 and 290/1=290.

Next, set the *Row offset* to 0 and the *Column offset* to 10 (i.e. the length of each interval). After pressing OK, the lines will appear as shown in Figure M42.

Figure M42

The next step is to use the *Extend* command to connect these lines to the longitudinal profile (Figure M43).

Figure M43

Now, since the profile is digital, the project station and elevation can be obtained by using the *Data Extraction* command to extract the coordinates of the end of the lines. This command is described in Chapter 3 and Exercise 73, so in the following, we will briefly explain the method.

After executing this command and completing its initial steps, use the *Select Objects* button to select all vertical lines. In the following steps, select only the *End X* and *End Y* options in the *Geometry* pane, and finally, save the file in .csv or .xls format. This file will be used to prepare the IDX files of the route.

During exporting, remember to check whether the profile points are the start or the end of the lines. Because if they are at the start of the lines, you should select the *Start X* and *Start Y* options from the *Geometry* pane.

101-Extraction of texts from a drawing

Sometimes, it is necessary to extract the texts of a drawing for use in other software such as Excel.

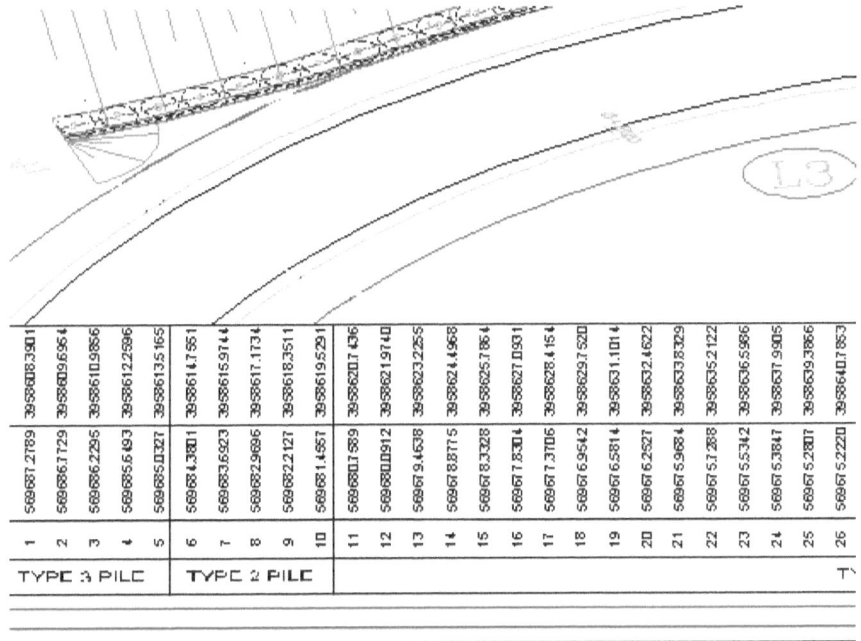

Figure M44

For example, the drawing illustrated in Figure M44 is a plot of a wall consisting of 53 piles, with the coordinates of every pile displayed at the bottom. The goal is to extract these coordinates for use in Excel.

This drawing is named <u>Pile.dwg</u> and can be found in the folder <u>dwg</u>.

Open this file, and then execute shortcut command *dx* in the command line to open the *Data Extraction* window. In the first page of this window, select a name for the file and click *Next*.

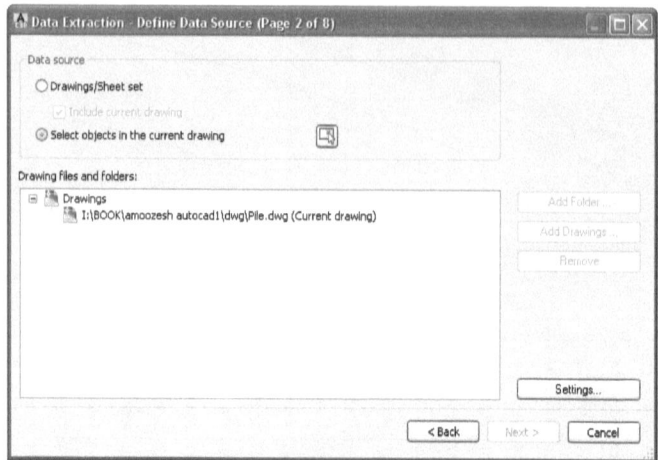

Figure M45

In the second page, choose *Select objects in the current drawing* (Figure M45), then use the button in front of it to select all the texts related to pile coordinates. After selecting all texts, press enter to reopen the window and then click *Next* to proceed to the next page

In this page, you will see a list of selected objects. As shown in Figure M46, check the *text* option and unchecked the rest.

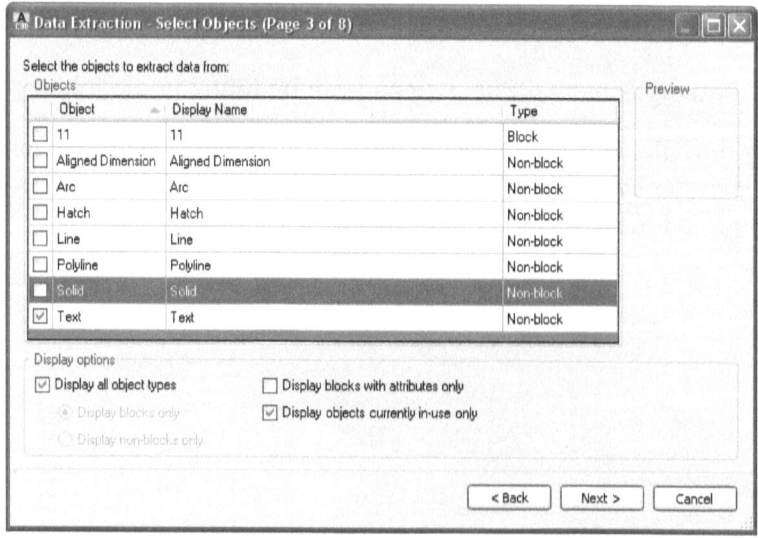

Figure M46

When done press Next to proceed to the fourth page of this window.

In the page, you should enable the *Text* option in the *Category filter* box and only the *Value* option in the *Properties* box (Figure M47). Press next when done.

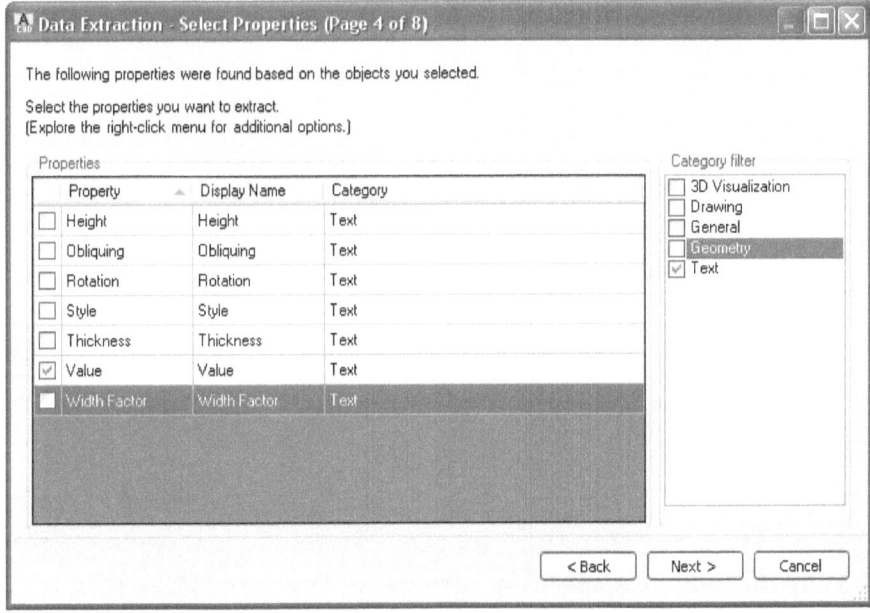

Figure M47

The next page will show you the selected texts. Click *Next* to go to the next page.

In this page, select the *Output data to external file* option and press the button below it to select a name and a path for the output file. When done, press *Next* in this page and *Finish* in the final page to finish the work. Now, the exported file can be opened with Excel, where you can perform the desired operations on the exported data

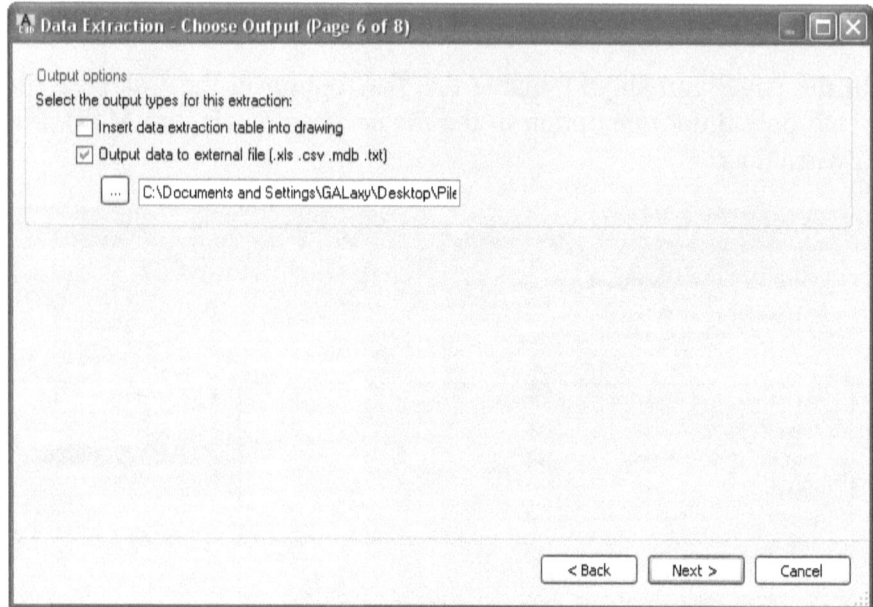

Figure M48

102-Why the software may stop displaying Open and Save menus and ask for the file path in the command line

Sometimes, when you try to save or open a drawing file, the software does not open the corresponding windows and instead asks you to enter the file name and path in the command line (for example see Figure M49).

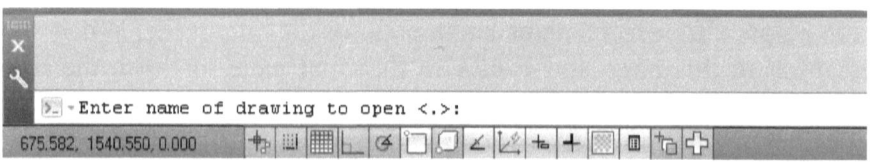

Figure M49

This is while most users prefer to open or save their files through the dedicated menus.

To resolve this issue, type the shortcut command *Filedia* in the command line and press enter. As shown in Figure M50, the software will then ask you to enter a value for the *Filedia* variable in the command line.

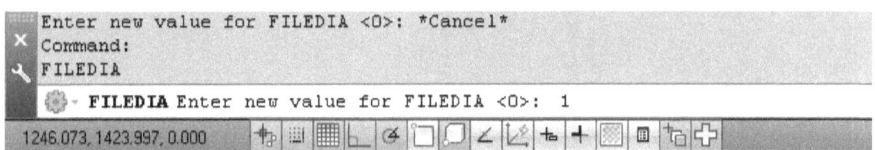

Figure M50

Filedia is one of the AutoCAD system variables that can be set to either 0 or 1.

When set to 0, this variable forces you to use the command line for saving and opening files, but setting this variable to 1 enables you to use dedicated menus for this purpose.

103-Why you may not be able to Zoom and Pan with the mouse wheel

By default, you can use mouse wheel as a replacement for Zoom & Pan commands. More specifically, you can roll the mouse wheel up to zoom in and roll it down to zoom out, and move the mouse while pressing the wheel button to pan around the drawing. But sometimes the mouse wheel does not perform these functions, and pressing the wheel button opens the *osnap* menu shown in Figure M51. To enable the pan function of the mouse wheel, execute the shortcut command *Mbuttonpan* in the command line.

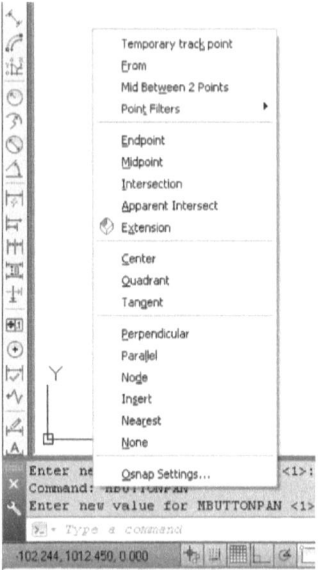

Figure M51

Mbuttonpan is also another AutoCAD system variable, which must be set to 1 to enable the pan function, otherwise, the software just opens the menu shown in the above figure

You can use another system variable called *Zoomwheel* to adjust the zoom function of the mouse wheel. When *Zoomwheel* is set to 0, rolling the mouse wheel up causes the view to zoom in and rolling it down causes the view to zoom out, and when it is set to 1, the wheel functions in the opposite direction.

104-Why sometimes *Join* command cannot be used to integrate two lines

It is not uncommon to come across two lines that cannot be integrated with *Join* and *Pedit* commands.

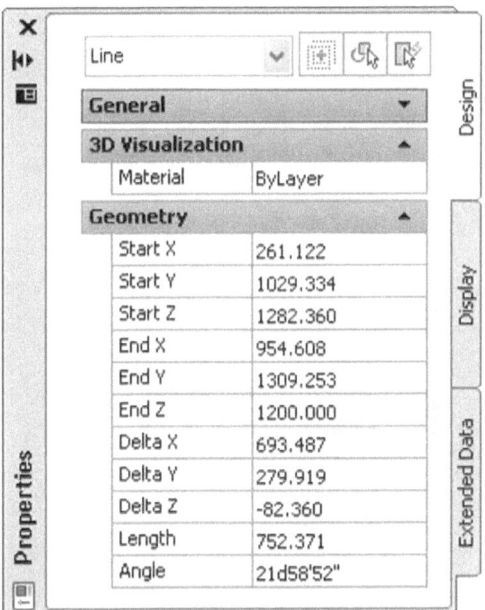

Figure M52

This problem may have different causes. For example, one of the lines could be a mistakenly created block. But the main reason for this problem is the difference between the elevations of the lines at the supposed point of intersection.

If you click on a line with explicit *Line* format and run the *PR* command to open the *Properties* window, you will see two parameters called *Start Z* and *End Z* (in the *Geometry* pane), which show respectively the elevation of the start point and the end point of that line.

In contrast, if you click on a *Polyline* and open the *Properties* window, you will find a parameter called *Elevation*, which shows the elevation of the entire line. The difference between these elevations is the main reason why sometimes two lines, two polylines, or a line and a polyline cannot be joined.

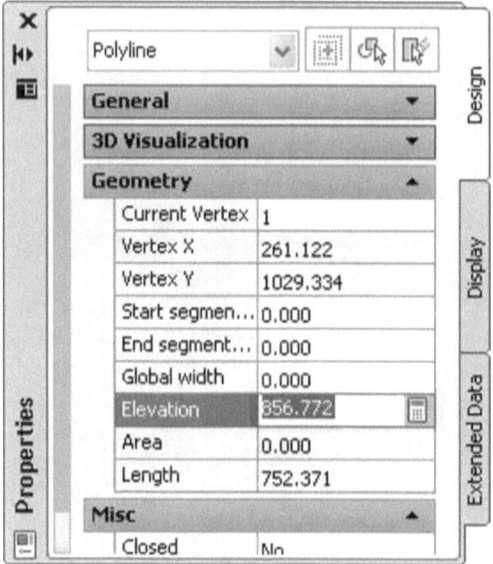

Figure M53

Hence, before integrating such lines, we must modify their elevations so that they match each other, albeit provided that their elevation difference is indeed small.

105-Why sometimes a layer cannot be deleted

Sometimes, when you delete a layer from the *Layers* window, a message like the one shown in Figure M54 informs you that the layer cannot be deleted. This can happen because of four reasons:

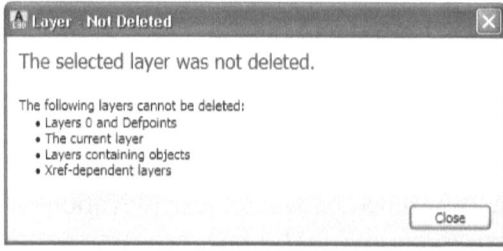

Figure M54

1- The selected layer is layer 0.

2- The selected layer is the current layer.

3- The selected layer contains object(s).

4- The selected layer is X-ref dependent.

If any of the above is true, AutoCAD will not be able to delete the selected layer. The best way to remove layers is to use the shortcut command *Laydel*. After executing this command, a message like the one shown in Figure M55 will be displayed on the command line.

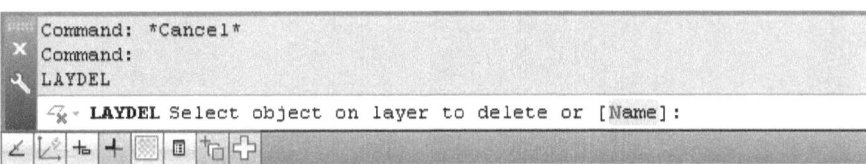

Figure M55

After viewing this message, you can delete that layer by clicking on one of its objects. You can also type *N* in the command line to view a list of all layers in the drawing (Figure M66), and then select the layer to be deleted.

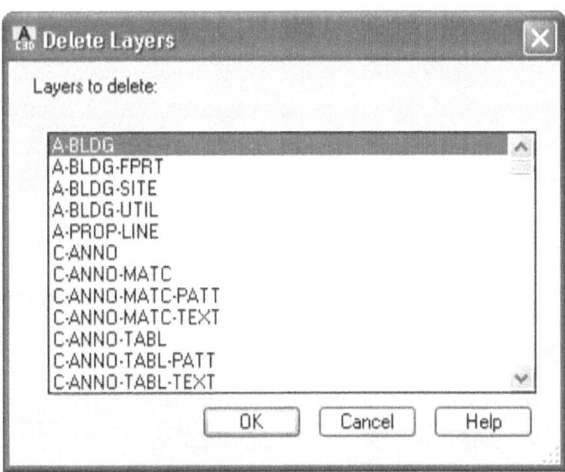

Figure M66

106-Why sometimes objects are missing from the plot

Often, objects do not appear in the printed plots because their plot mode is set to off.

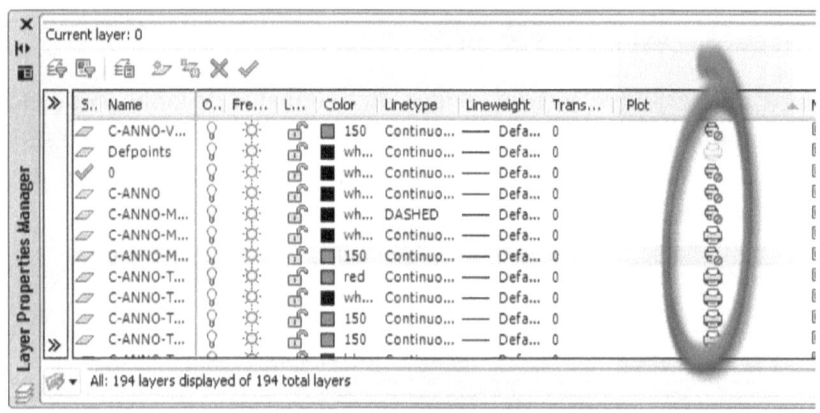

Figure M67

If you open the *Layers* window, you will see a column called *Plot*, where can decide which layers will appear in the printed plot.

107-Why sometimes zooming is too fast or too slow

You can use the *Zoomfactor* command to set how fast the view will zoom in or out when you roll the mouse wheel. This variable can be set to any value between 3 and 100 (3 for the slowest zooming and 100 for double zooming with each roll).

108-Why you may not be able to select an object first and then execute a command

For an operation such as erasing, you can first select an object and then execute the *Erase* command, or first execute the *Erase* command and then click on the object. The system variable that decides whether the first option is available is called *Pickfirst*. After running this command, you

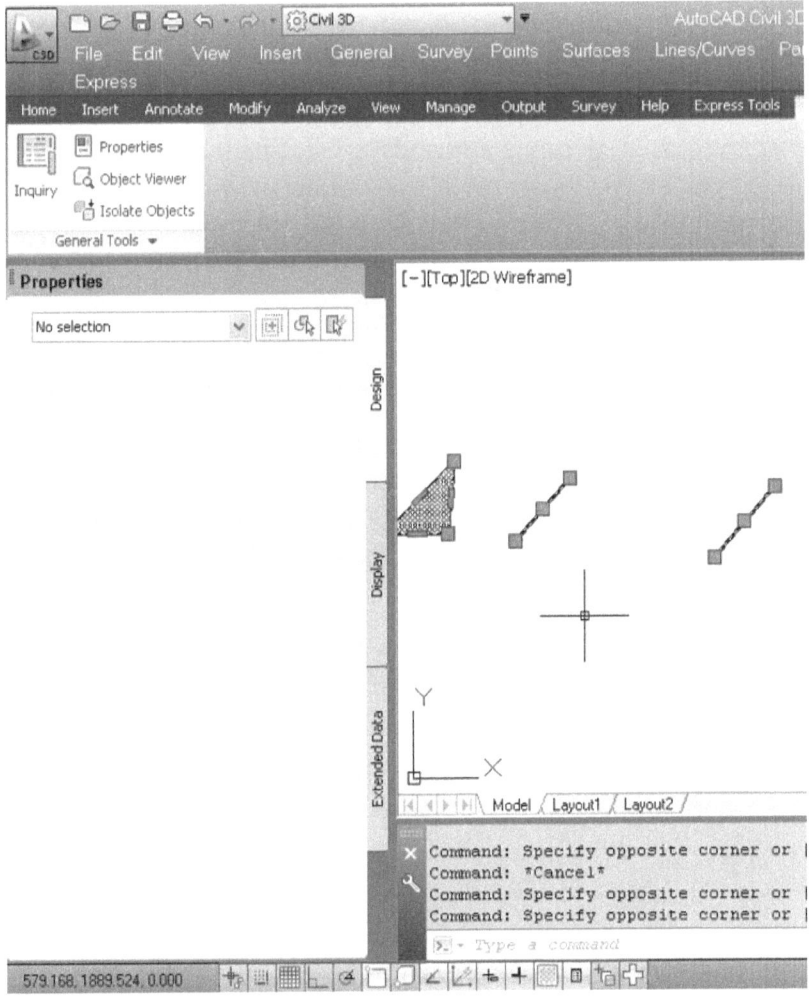

Figure M68

In AutoCAD, there is a variable called *Propobjlimit*, which determines the maximum number of objects whose properties can be changed simultaneously (it can take values between 0 and 32767).

In Figure M68, we have set this variable to 2, so the software cannot change the properties of three selected objects simultaneously.

can set its value to 0 or 1 to disable and enable the option of selecting the object first and then executing the command.

Note that disabling *Pickfirst* can prevent error, because when enabled, it allows the users to run commands on accidentally or unknowingly selected objects. In contrast, when this option is disabled, executing any command will unselect any currently selected object, thus forcing the user to specify the objects that will be affected by the command.

109- Why sometimes hatchworks are missing from the drawing

A system variable called *Fillmode* determines whether hatchworks appear in the drawing. After running the command *Fillmode*, you can set this variable to 0 or 1 to disable or enable all hatchworks in AutoCAD.

110-Why sometimes only one object can be selected by each mouse click

AutoCAD has a system variable called *Pickadd*, which when set to 0, prohibits each mouse click from selecting more than one object. To restore the normal mode of selection, execute this command and set it to 1.

111-Why sometimes the properties of objects cannot be changed from the *Properties* menu

In general, the *Properties* window allows us to change some properties of objects such as the layer, line thicknesses, line type, and even some geometric properties. But sometimes, AutoCAD does not allow us to change the properties from the *Properties* window and displays the message shown in Figure M68.